A Geographic Perspective of Cuban Landscapes

Landscape Series

Volume 15

Series Editors:

Aims and Scope

Springer's innovative Landscape Series is committed to publishing high-quality manuscripts that approach the concept of landscape from a broad range of perspectives. Encouraging contributions on theory development, as well as more applied studies, the series attracts outstanding research from the natural and social sciences, and from the humanities and the arts. It also provides a leading forum for publications from interdisciplinary and transdisciplinary teams.

Drawing on, and synthesising, this integrative approach the Springer Landscape Series aims to add new and innovative insights into the multidimensional nature of landscapes. Landscapes provide homes and livelihoods to diverse peoples; they house historic – and prehistoric – artefacts; and they comprise complex physical, chemical and biological systems. They are also shaped and governed by human societies who base their existence on the use of the natural resources; people enjoy the aesthetic qualities and recreational facilities of landscapes, and people design new landscapes.

As interested in identifying best practice as it is in progressing landscape theory, the Landscape Series particularly welcomes problem-solving approaches and contributions to landscape management and planning. The ultimate goal is to facilitate both the application of landscape research to practice, and the feedback from practice into research.

For further volumes:
http://www.springer.com/series/6211

Jennifer Gebelein

A Geographic Perspective of Cuban Landscapes

Springer

Jennifer Gebelein
Earth & Environment Department, ECS 467
Florida International University
Modesto A. Maidique Campus (MMC)
11200 S.W. 8th Street
Miami, FL 33199
USA
gebelein@fiu.edu

ISSN 1572-7742 e-ISSN 1875-1210
ISBN 978-94-007-2283-5 e-ISBN 978-94-007-2406-8
DOI 10.1007/978-94-007-2406-8
Springer Dordrecht Heidelberg London New York

Library of Congress Control Number: 2011940973

Printed on acid-free paper

Springer is part of Springer Science+Business Media (www.springer.com)

Preface

This book was inspired by my travels and fieldwork in Cuba over the past 8 years. The first time in 2003, I was invited to speak at a conference in Havana sponsored, in part by the University of Havana and its Geography Department. From the moment I landed until the time I left, I was overwhelmed by the generosity of the people, the impressive knowledge presented by scientists about the myriad of ecosystems in Cuba, and the beauty of the island. And as I returned to Miami from that first trip, I was struck that while Cuba seems so far away in our typical American mindset due to politics; it is so close in terms of geographic distance, that it seemed a travesty to *not* become involved in scientific research on the island. On subsequent trips my first impressions were only strengthened as I explored the island with my student. Upon traveling across the island to many different preserves, conservation areas and World Heritage Sites, it became clear that the management of Cuba's natural resources was a complicated and monumental task. I wanted to better understand how the Cubans administered these very different areas with distinctive conservation strategies. However I found that this administrative framework was exceptionally complicated and my only hope for grasping the full implications of current laws and policies was to go back in time and appreciate the many stages of landscape evolution that Cuba has undergone. This was the impetus for this book.

This book is the culmination of research that gives the reader a nonfiction view of how Cuba's landscape has changed since the time when Columbus first set foot on the island and encountered the Indigenous peoples who lived there in 1492 to present day. I touch on several topics that are relevant to analysis of landscape change over time to help the reader understand the full complexity of Cuba's physical transformation. That transformation from a heavily forested island to less than (currently) 18% forest cover is quite drastic and during more recent decades the government has established a system of protected areas and strong governmental controls over environmental policies and the manner with which the island can be built upon by foreign investors, urban expansion projects, or natural resource exploitation. This complex structure that has emerged is closely linked to improve environmental monitoring by scientists both in Cuba and abroad and many national and international collaborations to assess a best way forward regarding environmental policy. This book represents the first step in a long term project, so to that end it is wide ranging in terms of subject matter related to history, technology and geography. And while politics helped shaped the topics presented here, this is not a political book.

This book was completed by several trips to Cuba, many hours spent hiking, driving, informal interviews with Cubans in many welcoming homes during my fieldwork. The fieldwork itself involved long hours, generally extremely hot days and nights, and sore muscles from dragging heavy identification books, laptops and a cumbersome GPS unit through many hiking trails. It also required countless hours in many libraries retrieving out of print texts such as reprints of explorers' navigation logs and diaries, ancient atlases, and a constant hunt for digital GIS data as well as recreation of faded maps into new media.

There is no way to truly do justice to the many individuals who helped me with this enormous project. That help came in the form of helping me find my way when I was ridiculously lost in the field, kindly understanding that I really did *not* mean to stumble upon unmarked military land when traveling through remote areas. It also came in the form of wonderful colleagues who provided documentation of laws and policies and translations for them that I would have never found otherwise such as Dan Whittle of the Environmental Defense Fund (EDF) and Doug Radar also of EDF. I especially want to thank David Guggenheim for his continued belief in me and my work; he is a positive light in this world and makes an enormous effort to bring people together for collaborative work in Cuba. Fernando Bretos has also made a positive impact on me during this process and I appreciate his collaborative work as well. The person responsible for bringing me to Cuba for the first time is Michael McClain who was a wonderful mentor for my early Cuba work and I will always remember his infectious enthusiasm for research in Cuba. I would also like to thank the Cuban Research Institute (CRI) for funding my travel to Cuba. The anonymous reviewers made great improvements on this work and later comments and edits by colleagues were also very helpful. I would also like to thank Liz Greb and the Dean's Office in the College of Arts & Sciences for being wonderfully understanding particularly during the final six months of writing. To my former student Susanne thanks for being a terrific travel companion and translator. Finally, I would like to thank my family for their wonderful support and especially my husband Jamie who was amazingly supportive during this entire process.

Contents

Chapter 1
Historical Background of Cuban Land Cover Change

Abstract Chapter one introduces the concepts of landscape change, what defines a landscape, and how Cuba's land use practices have changed over time. Here the definition of land use is distinguished from land cover. The shifting perception regarding the island's natural resources is also described in terms of the various perspectives spanning from those of the first Indians to those of the early Spaniards. The question of what those varying perspectives meant in terms of exploitation versus

J. Gebelein, *A Geographic Perspective of Cuban Landscapes*, Landscape Series 15,
DOI 10.1007/978-94-007-2406-8_1,

conservation is also briefly explored. Finally, the chapter concludes with a look at how the landscape has changed, in map format, from the time of Columbus to present day.

Keywords Cuba • Land use change • Land cover change • Taino • Guanahatabeyes • Ciboneyes • Sugarcane • Landscape • Livestock • Mining • Environment

1.1 Introduction

Cuba has rich and complicated history. The result of dependence on its lands and natural resources for trade, commerce and survival is clearly demonstrated in the visual landscape we see today. Cuba's physical landscape includes examples of nearly every single environmental habitat found throughout the Caribbean islands. The microcosm of cloud forest showcasing the perfect habitat for orchids in Soroa; desert-like conditions along the coast east of Guantanamo; lush pine forests atop the Pico in the Sierra Maestra; and the Zapata wetland, a world heritage site, that provides refuge to endemic and migratory animals is one of the most significant wetlands in the Caribbean. These are only a few examples of the extremely heterogeneous landscape of Cuba. The island's physical characteristics that define its landscape have changed dramatically since the time of the Taino Indians.

The phrase "Cuban landscape" can have a myriad of meanings; however here the focus is on the **physical** landscape of Cuba. A physical landscape is a combination of an area's natural features and processes that help characterize and distinguish one area from another. An area's natural features include forested areas, soil types, mountains, rivers, valleys, locations of flora and fauna, for example. Processes that helped (and continue to help) shape these features include wind, rainfall, ocean and coastal currents, and air temperature, for instance. The way these elements combine to form a region or distinct area is how it is possible to distinguish one landscape from another. A physical landscape can be characterized as dry forest or savannah; wetlands versus rocky coastline; and all landscapes are defined by their unique physical properties that incorporate various types of defining elements.

The physical landscape has clearly not evolved in a vacuum; it is certainly *linked* to aspects of culture and politics throughout its history. And these links are tied closely to history as it has marched onward. These kinds of linkages will be addressed throughout this book during the various time frames discussed and described. The interpretation of what a physical landscape is has also changed when interpreted by Christopher Columbus in 1492, versus Diego de Velázquez in 1551, versus Jose Marti in 1895, versus Ernest Hemingway 1939–1960 (when he lived in Cuba), versus Fidel Castro in 1970. These different interpretations are so important because what something is defined as by any given observer; is tied closely to its *humanistic importance* in the bigger picture and thus, its true *meaning*.

This leads us to the interpretation of land use versus land cover. ***Land cover*** is defined here as how the landscape is quite literally covered by natural features. For example the land could be covered by forest, mangroves, shallow water like a wetland or marsh, bare soil, savannas, or grasslands. ***Land use*** is defined here as how the land is utilized by people. Examples of land use can be combined into categories such as urban, citrus plantations, suburban, cornfields, mining, landfill, or freshly tilled soil. There can be some overlap between the two landscape characterizations. Many times this overlap is the result of a fuzzy boundary between natural and man-made features such as parks adjacent to natural forested areas. There are different interpretations of land use but the overarching theme in land use is what people do to the physical landscape in short and long term time scales. For example, repeated slash and burn agricultural practices will generally deplete an area's soil nutrients over a much shorter time than carefully rotating crops in that same area and allowing the soil to remain fallow for a season. Completely deforesting a steep slope will greatly increase the soil erosion and runoff versus selectively logging the same steep slope. The latter choice of action will have a lower rate of erosion and runoff over a short *or* longer period of time. Thus, even at these early stages of anthropogenic impacts on the island, we can begin the narrative concerning the way the island and its resources are perceived, socially constructed and managed under the evolving modernity of Cuban society (Steinburg 2001).

The perception of landscape importance has changed since the Spanish conquest of Cuba. The reason perceptual importance matters is how we perceive the landscape through the lens of natural resource exploitation versus protection or conservation directly impacts our actions upon that landscape. Therefore, first we begin with the characteristics that make Cuba unique, then a description of the landscape in the early 1400s as it was first documented by Spanish explorers, and concurrently relate associated perceptions of the environment and how humankind has changed the landscape with time.

1.1.1 Uniqueness of Cuba

One of the things that make Cuba so unique is its geographical location and size. Cuba is actually an archipelago that includes the main island of Cuba and the Island of Youth (previously known as the Island of Pines). The Cuban Archipelago contains over 4,000 islands and the Isle of Cuba has just over 100,000 sq km, making it the largest Caribbean island. Its location in the Caribbean gives it a moderate tropical climate and its varying elevations and exposure to a multitude of different wind currents and moisture levels, in combination with the varied soil types throughout the island yield an example of almost every ecosystem type in the Greater and Lesser Antilles Islands. The topography of Cuba is also quite varied. For example, in the Santiago de Cuba province the eastern Sierra Maestra Mountains contain the highest peak in Cuba: the Pico Real del Turquino (in the Turquino National Park). At 1974 m high this park contains the pine forests of the species endemic to this area called the *Pinus maestrensis*. In this same high region the cooler temperate habitats provide a moist, shaded environment to species of ground dwelling ferns and epiphytes are also very common. This is in contrast to the western end of the island in the Pinar del Rio region, part of which is characterized by a flat landscape dotted with mogote-shaped hills which can reach heights of 500 m (Domech and Glean 2001). Mogotes are geomorphologic structures characteristic of Cuba and are remnants of eroded limestone sedimentary layers (Day 1978; Easterbrook 1999; Fairbridge 1968). This region is also characterized by caves which were inhabited by some of the earlier Indian tribes. These landscapes contrast with the vast wetland area on the south western coast of Cuba called the Zapata. The Zapata Peninsula is located on the southern and coastal side of the Matanzas province. The Zapata area is the most important and largest wetland in the Caribbean inhabited by hundreds of species of flora and fauna and another extensive cave system along its coast. There are also tropical desert conditions characterized by xerophytic vegetation such as succulent plants, *microphyllous* thorny shrubs and ephemeral herbs. The most widely distributed of all Cuban cacti is *Opuntia dillenii* and one of the more prevalent legumes in these conditions is *Erythroxylon minutifolium* (Seifriz 1943). One of the regions best known for this type of environment is at the far eastern end of the island near the town of Maisi, along the coast in the Guantanamo province. Another key visual cue in this region is the variety of small shrubs and trees that are dwarfed by constant winds. As the elevation and distance away from the coast increases, the plant community changes significantly as well. The Oriente mountain and valley region also supports many migratory bird species flying north and south towards various destinations. These are only a few prominent environments in Cuba and undoubtedly they have been impacted by anthropogenic influences throughout time; but these give the reader some insight into the huge variety of habitats that showcase Cuba's heterogeneity in terms of landscape types and features. More specific, location-based examples as well as regional snapshots of Cuba's landscape will be discussed and described in detail as we proceed through different time periods in the following chapters (Fig. 1.1).

1.2 Landscape Influence of Early Indians in Cuba

The original landscape of Cuba has been a subject of debate among historians. We begin with the first documented observations by explorers and historical discoveries by scientists. These observations contain descriptions of both the landscape and the indigenous people who lived there at that time; as well as reconstructive efforts of scientists to determine the composition of Cuba's original landscape. The importance of Cuba's land use by indigenous peoples at the time of the Spanish conquest of the island is relevant because it gives the reader an idea of the original state of the Cuban landscape. Certainly the Indians had an influence on landscape change, but their exact impact is unknown except for observations by early explorers to the region as well as archaeological findings since that time. However what *is* known and documented is described here. Indigenous land use was also tied closely to their perception of the land's importance in daily life and cultural habits and practices, thus a description of each major Indian group is given below.

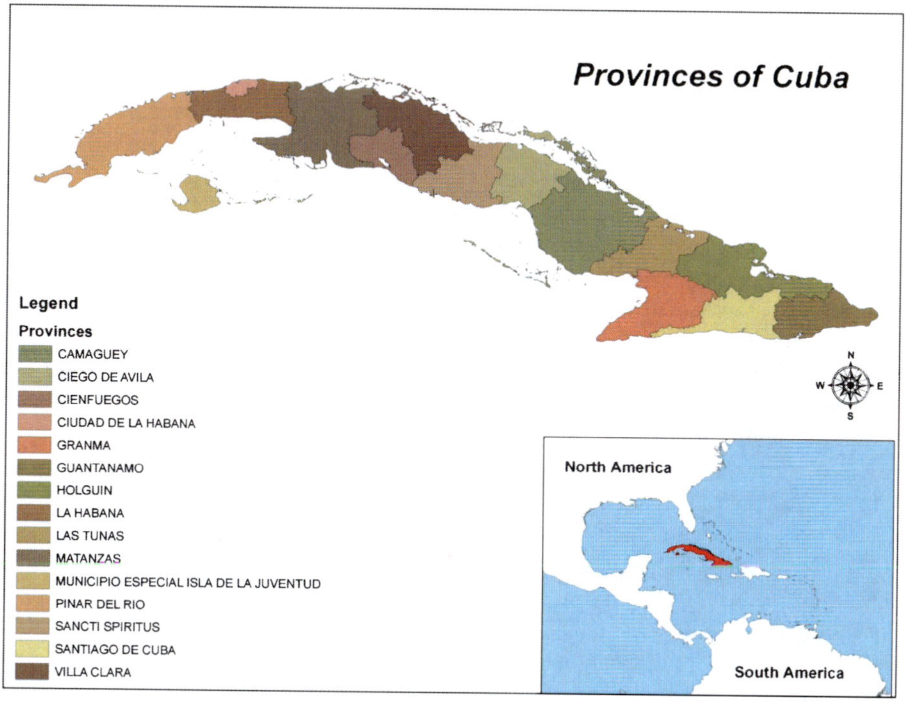

Fig. 1.1 Cuba's location in the Caribbean

Prior to the arrival of Columbus, there were three major groups: the Guanahatabeyes, Ciboneyes and Tainos. There was no written language for any of these groups and the most agreed upon total population at the time of Columbus' arrival is estimated at approximately 60,000 combined. However, this number is somewhat contested by the varied reports from different sources which range from 16,000 to 600,000 (Suchlicki 1974). The **Guanahatabeyes** did not build houses but lived predominantly in caves; by the time the Spaniards arrived they were largely living on the western end of Cuba and on the Peninsula de Guanahacabibes. From explorer accounts such as that given by Diego Velázquez (Cuba's first governor) in his letters to the King of Spain, these people lived mostly on forest fruits, fish and other animals such as turtles caught from the sea and other game captured inland. Archaeological remains suggest they also relied heavily on molluscs as an additional foodstuff. It is postulated that the Guanahatabeyes may have come from the southern part of the United States due to archaeological remains that exhibit a distinct likeness with relics of early Florida peoples. However there is a concurrent theory that they instead migrated from South America, through the West Indies and ultimately settling in Cuba. Since this early indigenous group subsisted mainly as a hunting and gathering society rather than an agricultural people their impact on the landscape was quite different than the Ciboneyes or the Tainos. Additionally, since the Guanahatabeyes lived primarily in caves, and did not build dwellings from forest materials such as the numerous pine or palm species. Overall they did not have a very large inhabitant footprint on the Cuban landscape. There is evidence of wood usage to fuel fires for cooking and comfort, but this impact on the Cuban forests was minimal. By the time the Europeans arrived in Cuba, their numbers were relatively small and by early accounts; their civilization and total populace appeared to be in decline (Suchlicki 1974); (Fewkes 1904). The **Ciboneyes** populated western Cuba and it is generally agreed that they were a subset of a larger Arawak group originally from South America. The Ciboneyes had a more diverse diet than the Guanahatabeyes that consisted partly of fish, birds, turtles and molluscs. They were reported to be highly competent fishermen and hunters and tended to reside in areas close to rivers or the coast sometimes in open campsites (Consculluela 1946). A minority of Ciboneyes occupied caves but most tended to lived in very basic dwellings termed bajareques or barbacoas. Their dwellings did not have much impact on their surrounding environment nor did their minimal agricultural practices according to Spanish explorers (Suchlicki 1974). The **Tainos** were also part an Arawak group who were present in Cuba approximately 200 years before the Spanish Conquest in 1511–1512. They inhabited eastern and central regions of Cuba as well as Puerto Rico Hispaniola and Jamaica. Archaeological investigations provide evidence that the Tainos either established permanent colonies or colonies that were temporary for the sole purpose of removal and export of locally plentiful resources. These two living patterns were typical of Taino culture throughout the Caribbean (Keegan 1995). When the Tainos remained in permanent settlements in Cuba they favoured areas near fresh water sources with fertile ground for agricultural practices. In these villages they established agricultural plots of cassava and maize which were principal food crops. The cultivation of yucca, a tuber, also played an important agricultural role, and with this particular crop they practiced cultivation in beds. Sweet potato, tobacco and cotton also became a significant component of the Taino market economy (Suchlicki 1974). To meet the demand for these goods, the Taino practiced slash and burn agriculture which meant clearing areas of forest and burning the trees and leaf litter to make the soil more fertile for crops. This type of agricultural practice has long terms impacts on the soil quality as well as increased erosion on hillsides where this custom sometimes occurs. Soil quality is important because the nutrients of the soil can be depleted from too much agricultural pressure on an area. This can directly dictate what types of plants can grow and thrive in that area in the future. If a formerly forested area, or areas dominated by shrubs and low trees, is cut to the ground and burned, over a period of time this can also cause a lack of vegetation growth whereby the soil has a much higher likelihood of washing away during the common torrential rainstorms experienced in the Caribbean (Schiettecatte et al. 2008). The lowland areas and some of the higher elevations were cultivated in this manner in the central and eastern portions of Cuba. The overall sections of cleared forest and shrubland were relatively low in comparison to the later environmental impacts the Spaniards had on the landscape. However, this began the vegetative transition from heavily forested terrain particularly in the coastal areas and lower elevations towards a landscape which eventually became agriculture. The Taino's environmental legacy was located mostly in the far eastern and to a lesser extent the central portion of Cuba. They generally did not extend their residence to western Cuba (Cosculluela 1946). Thus, as a baseline for comparison, the impacts of early Indians were relatively low. By early explorer accounts it is generally agreed that in the early sixteenth century (approximately 1500–1512) the estimated total forest land cover was between 88% and 92% of the island's surface. Within that category approximately 75–80% of those forests were classified as tropical forest whereas the remaining 25–20% consisted of pinewoods and short forests. The savannas did not occupy more than 3–6% of the Cuban archipelago (Monzote 2008; Rodriguez 1999).

1.3 Summary of Land Cover Change from Spanish Conquest to 1950

1.3.1 People's Impacts – 1950

It is well known that early maps created by cartographers had the most detail around coastal areas throughout the Caribbean and the world. This is due to the fact that the ship's captains had to navigate their vessels carefully over shoals, find inlets, safe harbours, and identify promising places to land their vessels along every coast. By comparison the interior areas of many regions were relatively unknown for many years unless there were rivers leading inland. It is important to note here that maps and nautical charts were the principal manner in which discoveries were sketched and shared with the Spanish Crown (Stevenson 1909). In the early sixteenth century a systematic cartographic historical record of all Spanish possessions in the Caribbean was begun starting with the establishment of the La Casa de Contratación (The House of Trade). This was a government agency which existed from the early sixteenth to and of the eighteenth centuries and operated under the rule of the Spanish Empire. It was established by Queen Isabella in Seville, Spain. It was overseen initially by Rodriguez de Fonseca, former administrator of the Columbus voyages. This trading house became a significant part of the Spanish crown's policy of imperial rule as well as directing commerce between Spain and its American possessions (Encyclopaedia Britannica 2011). It was abolished in 1790 but during those 200 years how the island of Cuba was depicted on the cartographic maps directly impacted how its resources were utilized and what marked the beginning of major land cover and land use changes on the island. The information for those maps came directly from explorers' accounts of the island, as well as their logs and diaries kept along their journeys.

The Spanish crown financed Christopher Columbus with three ships the Niña, the Pinta, and the Santa Maria to find a new water route to the treasures of the Indies, the islands located off the southeast coast of Asia. After months of travelling and despite a mutinous crew, on October 12, 1492 Columbus and his crew discovered the island of San Salvador in the Bahamas. Not long after this initial discovery,

Columbus found Santa Maria de la Concepcion (Rum Caye). Not long after this on October 28, 1492 Cuba was sighted. He sailed into Bariay Bay on the northeast region of Cuba. Columbus immediately claimed this land for Spain and for the following 5 weeks, sailed his ships along the north-eastern coast of Cuba vainly searching for gold (Suchlicki 1974; Crooker and Pavlovic 2010). On one of Columbus' landings during his first journey to Cuba, he described what he saw in relation to its meaning for the Spanish crown.

> ...and I climbed to a small mount to discover some portion of the land and he was not able to see anything because of the large woods that were fresh and full of fragrance... Then the shipmen gave voices saying they saw pines. I looked to the sierra and saw them so large and marvellous that I could not estimate their height and straight shapes that were as stout and thin as spindles. Then we realized we could build ships and an infinite number of decks and sheatings, and masts for the largest ships of Spain. I saw oaks, and a fine river, and riggings to make water-powered saws... And all the sierras were full of pines and all over the place there were diverse and beautiful timberlands... And in that same fashion there were valleys, just liked the mountains, full of tall trees that were a glory to see and pine forests that seemed to be plentiful. (Rodriguez 1999)

The arrival of the first European explorers did not *immediately* have a substantial impact on the landscape of Cuba. The main reason for this was simply that Cuba was part of an exploration of the Caribbean and one stop of many land claims for the Spanish crown. Therefore the purpose of these initial voyages was not for the installation of infrastructure, ports or urban areas, but to instead scout out and claim lands that would be ideal for that infrastructure at a later time. This is an important distinction that this first voyage of Columbus was one with the purpose of discovery, not to Christianize or otherwise exploit the island's inhabitants (Adams 1892). The initial permanent Spanish settlement in the Caribbean was established in Santo Domingo, a large port town on the island of Hispaniola, not Cuba. The first ruler there was Nicolas de Ovando in 1498 and he set the standard of forcing local Indians to work the land and become slave labour for the Spanish. This model of forced labour eventually became the labour model for Cuba as well. This is relevant to land cover change because it impacted what labour type the Spanish used to till the land and partly their inability to expand more quickly. It was not until late 1510 that a voyage was made from Hispaniola to Cuba. Therefore, during the time period from Cuba's first discovery by Columbus in 1492, to its first real settlement in Baracoa in 1511, there was no quantifiable damage to the natural land cover of the island. This was because with the exception of firewood for fuel and an occasional vessel, there was little felling of trees in the vast Cuban forests.

Christopher Columbus' son, Diego Columbus, was appointed governor of the Indies and lived in Hispaniola during this time. His major concern was expanding the territory under his rule. The first opportunity to address this goal was lead by Diego Velázquez, a lieutenant to the Governor of Hispaniola (Nicolas de Ovando). He took three ships and an army of three hundred European men with him to settle Cuba. The first settlement was at Baracoa (from 1511 to 1512), which he named Nuestra Senora de la Asuncion. The following six cities were subsequently established in Cuba: Bayamo in 1513, Puerto Principe (Camagüey), Sancti Spiritus and

Trinidad in 1514, and Santiago de Cuba and Havana in 1515. From this point in history until the mid-1700s most new communities were founded near Havana although Velázquez chose Santiago de Cuba as Cuba's first capital due to the fact that it had an excellent harbour and was close to the major trade routes at that time. Other areas of settlement during this time period included towns in Pinar del Rio, Holguin and Santa Clara. During this first period of colonization the main developments regarding natural land cover were those areas listed above and their associated ports becoming less forested as town sizes slowly increased and the need for lumber also increased. Additionally, it was not only the forested lands along and slightly inland from the coastal areas that were reduced, but also some mangrove areas as well. However this population and infrastructure expansion was hindered by several factors. The first was the departure of many of the original Spanish settlers on Cuba for the American continent, lured away by its conquest and perceived new opportunities (Monzote 2008). The second was the Indian resistance. From the moment Velázquez landed in Baracoa he and his men encountered hostility from the Indian population. The major source of this came from an Indian chieftain named Hatuey, who was originally from Hispaniola and had witnessed the enslavement and brutality Velázquez had used to subjugate the Indian population in Hispaniola. The resistance led by Hatuey did not last long; he was eventually captured and burned at the stake. However some resistance did continue in the form of ambushes as the Spanish sought to expand their lands to the interior of the island. These types of attacks were quelled with the introduction of an agreement that convinced the Indians to put down their weapons and work close to the new towns that were established on the island, and work *for* the Spanish settlers (Suchlicki 1974). Over time the Indian population diminished by assimilation into the settlers' population, sickness brought by the settlers, escape from the island, death due to enslavement practices, or from revolts.

The early development of towns into cities during this early period was seen especially in Havana where development spread out in a radial pattern away from the city centre by the port. It was in 1717 when the harvesting of lumber for shipbuilding was easily observed from the forests around Havana. This spread to more coastal deforestation with greater and greater intensity until in 1748 regulations were disseminated to conserve these areas for the navy. In short, this meant that forests 25 leagues distant from all coasts and navigable rivers now fell under the jurisdiction of the Spanish government. This total area was approximately one-half to two-thirds of Spain's entire woodlands (Monzote 2008). This was a turning point for the forests and for the perception of those forests in Cuba's history that will be discussed later.

1.3.2 Livestock Impacts – 1950

The long term goal of the Spanish crown was to establish land ownership structure that would support large property sizes specifically for raising livestock and farms for agriculture. This structure came in the early 1500s when the town councils in the first cities commenced allocating land by means of *mercedes* (land grants), which was a system of granting enormous properties that subsequently became known as *corrales* and *hatos* (small and large) (Thomas 1998). These land grants were in the peculiar shape of circles, whereby the owner was given land to a certain radius from a specific point. The *hatos* were 22,512 ha in size versus the *corrales* which only contained 5,628 ha (Monzote 2008). The *hato* were to be used for raising cattle and the *corrales* were to be used for raising pigs. Inclusively these terms were all considered a part of a proprietor's *hacienda*. The general agreement between the proprietor and the Spanish crown was that the proprietor acquired the usufruct of his land with the condition that he provided the nearby town inhabitants with beef and provided an inn at the centre of the property. There were many issues with this type of property delineation because the individual property lines were usually quite vague, and since the haciendas were circles, they tended to overlap and cause disputes among land owners as the island became more populated (Thomas 1998). During the time of the first established towns, Cuban forests suffered little damage except for fuel for firewood and building materials for the occasional vessel. On these circular land plots were growing numbers of cattle and pigs that were allowed to roam freely. The ranching by and large took place in the more level flatlands and rolling hill country of Cuba where there were less dense forests. Thus the damage by herds was generally to the lower branches and juvenile trees. Additionally, during this initial colonization period, the cattle were also placed on land that was a mixture of less dense forests with open savannas. The ranchers discovered that to support cattle on a large scale there was a need to move them during the different seasons of the year. For example, between October and March, which is the dry season, the grasses in the pastures diminished considerably and thus the livestock were moved to forested areas and ate low branches and leaves, fruit that had fallen on the ground, and reeds near the rivers (Monzote 2008). Raising livestock continued to be a profitable endeavour on Cuba but as the need for land to grow cane increased, some of this grazing land was converted to sugarcane fields. However, not all cattle grazing lands were converted to sugarcane fields. In 1860 there were still 6,000 cattle farms and 34,000 smaller livestock farms (Thomas 1998). Pertaining to the grasses that fed these livestock, there were several grasses introduced by ranchers for cattle consumption that remain there today. The reason for their introduction is the native grasses were viewed as suitable for grazing livestock *however* the cattle would subsist, not grow fat when eating these grasses. Thus two types of grasses were introduced to Cuba for this express purpose: The Parana grass from South America which grows well in lowland areas, and Guinea grass from the west coast of Africa. This second type of grass adapted well to steeper hillsides as well as higher elevations. The third type of imported grass for grazing was Bermuda grass imported from the United States, this tended to be the favourite of most livestock. Since these three types of grass were introduced they clearly changed the natural landscape with their successful spread in grazing regions all across the island, replacing native grasses in many areas while not entirely driving it out of existence. And cattle ranching continued to be a profitable venture into the 1900s partly due to the success and rapid spread of these non-native grasses. For example in 1895 over a million cattle were registered in the Province of Camagüey alone. They were fed almost exclusively on the Guinea and Parana grasses, plus fruits which dropped to the ground such as mangoes (Johnson 1920). By the early 1900s most cattle ranches were in the eastern half of the island, particularly Camagüey's cattle district. There were 5.3 million head of cattle in 1920, 5.0 million in 1930 and approximately 4.5 million in 1955 (Thomas 1998).

1.3.3 Precious Wood Impacts – 1950

Apart from herd damage there was some removal of forest tree species to provide the Spanish crown with precious woods such as the sabicú (*Lysiloma sabicu*), cedars (*Cedrela odorata*), and Cuban mahoganies (*Swietenia mahagoni*). According to Rodríguez (1999) the removal of these precious woods accounted for hundreds of cubic meters between approximately 1572 and 1590 for the construction of the convent of San Lorenzo del Escorial. There was also minimal clearing for small farming plots, and slowly expanding the space needed for settlements (Rodríguez 1999). Before the industries of tobacco and sugar, there was the industry of cutting down and exporting precious woods such as ácana, quiebrahacha, mahogany, and ebony for both the Spanish crown, pirates and smugglers (Monzote 2008). Overall, the colonization of the island, livestock ranching and deforestation that occurred for production of lumber had a fairly small impact on the amount of forest cover lost by the close of the seventeenth century, with approximately 90% of the forests remained intact (~5 million hectares) (Díaz-Briquets and Pérez-López 2000). However after the seventeenth century the demand for these types of wood increased partly due to better access and a higher intensity in logging efforts for ship building. Throughout the nineteenth century and the early twentieth, Cuba exported a substantial quantity of precious woods and general lumber. This was due, in part, to private citizens cutting down forests on their own property to help defray investment costs for enormous financial investments in growing, harvesting and processing sugar (Monzote 2008).

1.3.4 Tobacco Impacts – 1950

One of the crops that were initially cultivated in Cuba was tobacco. In fact Columbus and his men first saw the Cuban Indians smoking tobacco near what is now Holguín, thus tobacco existed as a crop on Cuba before the Spanish arrived and recognized its potential. At first, the Spanish grew tobacco along the banks of rivers where it was easily accessible, near the cities of Havana and Trinidad for example. This cultivation led to clearing of grasses and tropical forest near the river edges. There was much contention during this early period between the tobacco growers and the cattle ranchers because the livestock, which ranged wild, were allowed to trample the delicate tobacco plants. Tobacco grew in importance and environmental impact during this period because of its increasing popularity, especially with pirates and smugglers who would come to Cuba to acquire the leaves and also in the form of snuff (Thomas 1998). Small scale tobacco farming had clearly been taking place with minimal impact since before European colonization. However even after Spanish settlers came to Cuba and established legal and illegal tobacco trade, tobacco did not add much to Cuban commercial production because there was simply not much demand for it in comparison to raising livestock. The real demand for tobacco began in the early 1600s with the first shipments of tobacco registered in Havana in 1626, although history shows this was the first *formally registered* cargo, but certainly not the first Cuban tobacco to make its way out of Cuba across the Atlantic. By the mid-seventeenth century the export of tobacco from Cuba was a regular occurrence and was extremely popular in Europe, particularly in Spain (Ortiz 1947). By the middle of the eighteenth century tobacco was, besides sugar, one of the most valuable exports for the Cuban economy. Tobacco was grown in several specific areas in Cuba, according to (mostly) soil quality and rainfall. For example while the tobacco of the Havana Province was considered exceptional, it was still inferior to the tobacco cultivated in Pinar del Rio. The rule of thumb for tobacco growers was the further westward, the better the tobacco. In line with this rule, the tobacco industry recognizes four major types of tobacco. In the far east in the province of the Oriente is the *Mayari y Gibara* tobacco. Moving westward the *Remedios* in the Santa Clara province and the *Portidos* in Havana are of superior quality. And in the far western end of the island, in the Pinar del Rio province, in an area 90 miles long and 10 miles wide, immediately south of the Sierra de los Órganos is the location that holds the world's finest tobacco: the Vuelta Abajo district. Its success is attributed to the soil that consists of the Greenville, Ruston and Orangeburg fine sandy loams that have reddish, friable sandy clay subsoil (Whitbeck 1922 and Bennett 1928). Tobacco is generally grown on flat ground and is a fairly labour intensive to cultivate. One must place a light cloth over the furrows in the ground where the seeds reside, to prevent excessive light and heat from damaging the seed, in addition to many minute steps to cultivate fine tobacco. There are many complex policies, restrictions, revolts and politics that impacted the growing and selling of tobacco but here we are focusing primarily on the landscape history of tobacco. Tobacco's impact on the environment was certainly not as damaging as the growing and processing of sugar, which will be discussed shortly, but eventually an awareness of land degradation due to the overuse of soils from tobacco and sugar planting emerged. This came from direct observations by the farmers of their lands, which had been consistently slashed and burned over generations of tobacco plantings, that the nutrients of the soil were being depleted. The second observation was the result of erosion from clearing large tracts of forestlands for plantings. Over time erosion became more and more of an issue the more land was cleared, particularly with the massive rainstorm events which swept over Cuba every year during the rainy season. The results of the early observations led to soil and land conservation policies which will be discussed in Chap. 2. But by 1950, these two issues alone, as a result of deforestation for intensive tobacco and sugar farming, had changed the landscape forever from dense forest cover in 1492 to many large open areas of tobacco farms by 1950.

1.3.5 Sugar Impacts – 1950

One of the major *later* causes of deforestation was the cultivation of sugar. Even though sugarcane was introduced on the island by Governor Diego Velázquez in the early 1500s, it was not until 1590 that sugarcane was grown for industrial purposes using fairly primitive sugarcane mills. Sugar had a slow growth for the next several decades. The economic potential was recognized by farmers however it was also perceived as a noxious industry. For example in the first half of the seventeenth century local residents of Havana raised formal complaints about the destructive nature of sugar farming (Rodríguez 1999). The complaints were specifically aimed towards the large tracts of forest which were slashed and burned to build sugar mills in the area near Havana by the La Chorrera dam. Additionally the residents claimed that the refuse left over from the processing of the cane such as vast amounts of lye, ashes and waste from the use of 50 horses and mules was resulting in destruction of quality of life and reduction in forest cover (Marrero 1975). Consequently even at this very early date the negative environmental impacts of sugarcane processing were felt. However the negative impacts on the forests did not ultimately halt sugar's immense upswing in production that was to come because the damage to the forests and the refuse from the processing was fairly localized. Despite sugar's growing popularity, the entire process from clearing forest, planting, maintaining, processing and finally arriving at the final product was extremely labour intensive and required the use of many hands. Because, at first, Cuba did not have ready access to the labour it required, this produced a greater number of smaller mills with a lower output until several events occurred that changed the landscape forever. First, the British captured Havana in 1762 until 1763 (when the Treaty of Paris returned Havana to Spain). With this capture came the introduction of thousands of African slaves which proved to be the labour that Cuba had needed to support the production of sugar. Secondly, in 1778 the Spanish regulations and tariffs on trade restrictions were diminished considerably. And third, the Haitian Revolution resulted in a departure of Cuba's major competitor in the sugar export market. This revolution was a slave revolt which occurred in 1791 to obtain human rights and eliminate slavery (Rodríguez 1999; Thomas 1998; Suchlicki 1974). With these three factors alone the success of sugar and the downfall of the Cuban forests began in earnest. Historic records estimated that from 1775 to 1827 approximately 1.5 million hectares of forests were cut down and almost 60% of this total was directly linked to agriculture and the sugar industry. In 1827, 7.5 million hectares of forests (68% of the island's territory) still remained. Also in 1827, a quarter of all sugar exported from Cuba came from the province of Matanzas (Díaz-Briquets and Pérez-López 2000). Between 1828 and 1900 the total estimated area of deforestation was about 2.9 million hectares (Rodríguez 1999) and at this time the production of sugar was concentrated in a wide band extending from the province of Havana to the city of Cienfuegoes in the middle of the island. During the majority of the nineteenth century, 90% of all Cuban sugar exported originated from central and western provinces (Díaz-Briquets and Pérez-López 2000). In 1900 there were still approximately 4.0 million hectares of forest left, which was about 50% of the original forest covered estimated to be on the island in 1492. At that time, the majority of the island's western natural land cover had had its forests destroyed or irreversibly degraded largely due to the Cuban sugar expansion. At that moment sugar production was expanding eastward predominantly into the provinces of Camagüey and the Oriente. Regarding the devastation of Cuban natural lands, this expansion ultimately destroyed approximately 1.3 million hectares or about 102,313 ha per year from 1901 to 1913. By 1919, the areas that had either undamaged or minimally damaged ecosystems and habitats were found in the mountainous areas in the Sierra Maestra (to the east), the Zapata and Guanahacabibes peninsulas, the plains region to the south of Camaguey and Ciego de Avila, the central-western region of eastern Cuba (including the Cauto valley) and the southern half of the Isle of Pines. Between 1920 and 1926 the amount of total forest cover was barely 2.2 million hectares, which was only 20% of Cuba's total landscape. During roughly this same time frame, the annual deforestation rate was just over 75,000 ha. Between 1927 and 1958 the level of landscape destruction slowed considerably as a consequence of the quota system on sugar set by the United States, however there was still an overall decline of forested areas of 424,882 ha with an annual natural (forest) habitat loss of 13,277 ha. Thus in 1958 there was a total of 1.8 million hectares remaining of natural areas (particularly forests), or 16% of Cuba's total land cover. While this may seem like a fairly large extent of natural areas remaining after colonization, widespread crop introduction and a relatively high island population growth, the trees that were left were either the least desirable species or trees that were located in areas that were prohibitive to agricultural or urban development (Rodríguez 1999).

For clarification, there are several reasons why the sugar industry was more destructive to Cuba's landscape than any other industry. The first is that large areas of forest and shrub lands were slashed and burned, the ashes acting an initial fertilizer. The areas were cleared for space to plant the cane, and the lumber was used for the buildings supporting the mills (slave quarters, boiling house, slave hospital, sheds, stables, carpenter's shop, et cetera), the mills themselves and fuel for fire. The need for firewood was so crucial to the mills that forests directly surrounding the mills were soon devoid of trees and the labourers had to search further and further away for wood to burn. The mills had a lifespan of 40 years at best. Along with two or three sowings of canes which lasted 15–20 years each, the soil could provide no more profitable harvest. At that time, the farmer would either close down the mill and/or move onto other unexploited lands (Thomas 1998). The result of this was soil erosion would occur in many areas, particularly in locations with a gentle to steeper slope. As mentioned earlier, Cuba is subject to torrential rainstorms common in the Caribbean during the rainy season (July to October). The importance of this is once the landscape is denuded and the soil has no nutrients to support species of plants with a root system to hold the soil, it is easily washed away which is a serious problem for an island. Millions of hectares of forests and other natural lands were changed in this manner due to the development of cane sugar alone. Clearly other crops impacted the natural landscape as well, but the development of the sugar industry is clearly the most damaging in terms of land use and land cover.

1.3.6 Other Impacts

Although sugar cultivation and processing, tobacco farming, cattle grazing, and extracting precious woods were undoubtedly to blame for the majority of land cover change and degradation in Cuba from the time of Columbus to present day, there were other factors that should be mentioned. Urban expansion to accommodate a growing population, particularly along coastal areas, was a steadily increasing factor in the razing of natural areas for building structures and dwellings. Commercial lumbering was also a factor that steadily increased; and not simply for precious woods but for export and domestic utilization as well. In terms of monetary exports, between 1899 and 1927, wood exports ranged between one and two million dollars on an annual basis. This logging trend continued into the 1930s and 1940s. In addition to this, there were also open-pit mining operations developed in north-eastern Cuba which contributed to large-scale deforestation in the 1940s (Reed 1992). Another practice from the time of Columbus until this period was continued slash and burn agriculture by farmers who did not own land nor have access privileges to lease lands. These were subsistence farmers and this practice was done mostly in the Oriente Province Mountains. Interestingly, during the 1940s and 1950s the principal cause of forest loss was the use of wood as fuel. During this time period, almost 70% of all domestic lumber was turned into charcoal or utilized directly as fuel (Marrero 1950). Therefore between the colonization of Cuba and 1950, not only were there were the major land uses that degraded the natural cover, but there were also these additional smaller factors that also had an impact on Cuba's landscape.

1.4 Evolution of Land Cover from 1950 to Present Day

The reason for choosing 1950 as an evolution breakpoint is this was the beginning of a turbulent time in Cuban history regarding the physical landscape but also the cultural and political landscapes as well. Since these are all closely related it is important to draw this line in time moving forward in its own era of transformation. The second reason for this subsequent timeframe is a decade later the space race began. This is the technology that laid the groundwork for observing the earth from space, including landscape changes. Third, beginning in the 1950s there began to be serious government policy attempts to control the deforestation, soil erosion and nutrient depletion of soils due to decades of poor agricultural practices. Thus it is important to separate this second era from the first to focus on those three main thrusts, which will be addressed more in later chapters.

In this section the landscape *changes* that took place from 1950 to present day will be described with a brief explanation regarding why the changes occurred. However the major focus on policy changes affecting environmental change will be discussed in greater depth in Chaps. 2 and 3. The land cover and land use categories will be combined into one section here.

In the 1950s there were several major land cover and land use categories which can be clearly identified for the whole island. In terms of *land use* types these were urban, livestock grazing and agricultural areas. Beginning in the 1950s through present day there have been increases and fluctuations of these land use categories. Historically, cattle-rearing has a longer history than sugar. The majority of the ranches existed in the eastern half of the island, especially the cattle district in Camaguey. Interestingly, there were roughly the same numbers of cattle in the 1950s as in the 1920s (4.5 million in 1955 versus 5.3 million in 1920). In 1959 pastureland was available on the natural grasslands that occupied between 35% and 45% of the entire island at this point in time. This resulted in Cuba becoming an importer of beef rather than a large exporter as it had been in the past (Thomas 1998). Urban areas were expanding with an increasing population. Sugar exports collapsed due to the fall of the Soviet Union in 1990. As noted by the Comisíon de Conservación de los Recursos Naturales (CCRN) the forests that still existed in 1962 had predominantly lost their capacity to yield some of Cuba's most valuable woods because of continued selective harvesting (Grupo Cubano 1963). Tobacco continues to be grown and exported as a luxury item with particular demand for the tobacco grown in the western half of Cuba. Land cover has also experienced some interesting changes since 1950. There has been a noted increase in grasslands and shrub lands as agricultural fields are left to become fallow, allowing for succession grasses and then later, small shrubs and trees to grow. This thereby supports the idea that an increasing number of opportunistic vegetation types have invaded cleared areas. These types of trees and plants appear as a result of the deterioration of natural vegetation types and communities, particularly in formerly forested areas. The kapok tree (*Ceiba pentandra*) and royal palm (*Roystonea region*) are both examples of such opportunistic trees (Domech and Glean 2001).

In 1945 Díaz-Briquets and Pérez-López (2000) estimate that **forests** were 18–20% of the national territory. This decrease was shown to continue when in 1992 Cuban officials presented an estimate at the United Nations Conference on Environment and Development that forest comprised only 14% in 1959. From these figures we can estimate that during the time period from 1945 to 1959 Cuba lost 460,000–800,000 ha of forest (Díaz-Briquets and Pérez-López 2000). Shortly after Castro took power he instituted a reforestation program. This was a win-win situation since it served the political goals of improving rural employment and socioeconomic opportunities and national development priorities, especially in mountainous regions. According to official estimates, the overall result was positive and by 1992 18.2% of Cuba's landscape was classified as forest. Of this 18.2%, 84% was

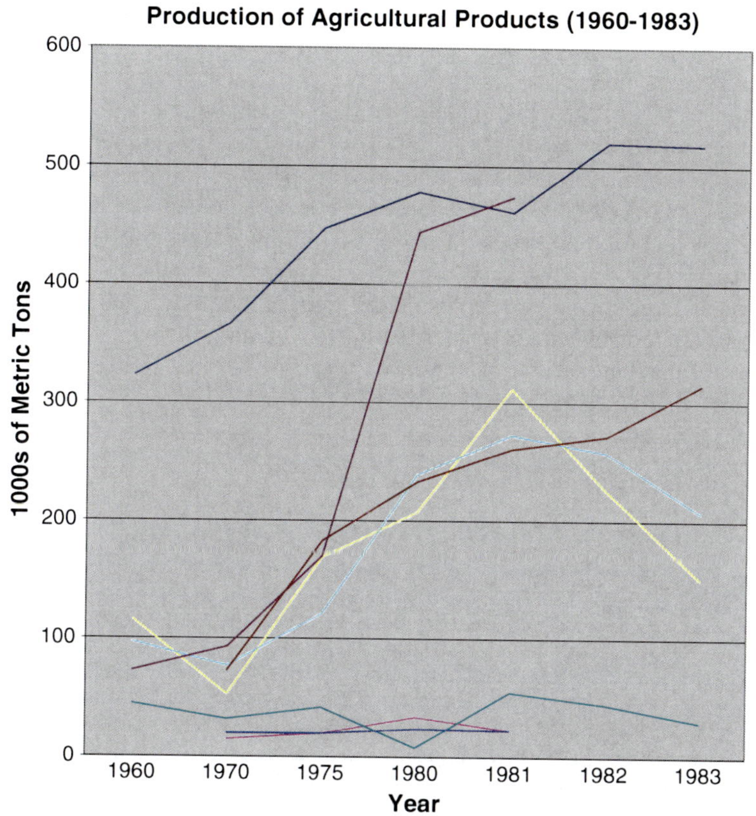

	1960	1970	1975	1980	1981	1982	1983
Rice	323	366	447	478	461	520	518
Corn		15	20	34	23		
Tomatoes	116	53	169	207	312	226	151
Potatoes	97	77	121	239	273	258	207
Citrus	73	93	170	444	474		
Bananas		73	183	233	261	272	315
Tobacco	45	32	42	8	55	45	30
Coffee		20	20	24	23		

Rice Corn Tomatoes Potatoes
Citrus Bananas Tobacco Coffee

Graph 1.1 Showing growth of agricultural products from 1960 to 1983 (Data source: Economia, Poblacion y Territorio en Cuba (1899–1983))

considered natural forest and comprised two million hectares. Currently, natural forest cover is estimated between 10% and 14%, depending on the source.

The production of other agricultural products besides sugar increased significantly from 1960 to 1983 (Graph 1.1 – sugar not included). This is most obvious in the increase of production of rice, which is one of the main staples in the Caribbean (Luzon 1987). In the same time period the production of bananas has more than quadrupled, this may be due to banana production exports to East European countries such as the German Democratic Republic (GDR), Poland, Hungary, and to the Soviet Union.

Sugar is one of the driving factors behind the changes to Cuba's landscape and its economy. Regarding the economy sugar was the monoculture upon which Cuba largely based its trade relations with the Soviet bloc countries. With the fall of the Soviet Union in 1990 came the end of the trade relationship between Cuba and the Soviet Bloc countries as well. This led to a drop of 47% of metric tons of sugar output in just 4 years. Cuba was not able to replace lost markets, or, even if it had found new foreign buyers for its sugar product, it did not have the currency to purchase supporting machinery, fuel, pesticides and fertilizer for its crops. The impact this had on the landscape was the severe drop in cultivation of sugarcane and focus on other types of subsistence crops. Cuba had to focus inward rather than seek help from other nations. This is discussed in later chapters.

1.5 Where Was the Greatest and Least Change?

1.5.1 An Acknowledgement to Past Classification Efforts

This is not the first effort to illustrate land cover categories or define certain areas of natural biodiversity, urban growth, deforestation, or some sort of change over time. There have been several other notable efforts in this regard that are worth mentioning here. Of course some of the early explorers documented their surveys extraordinarily well such as Columbus' "Navigation Diaries" as well as much later; Alexander Humboldt in the 1800s. Humboldt (1856), in his "The Island of Cuba" describes in detail the island's geology, fresh water locations and quality, the varied soil types, the climate, the geography of the reefs around Cuba, as well as other topics. Some of the original work on organizing the plant life of Cuba was done by Hermano León who published two articles of note on this topic: "Flora de Cuba" (1926) and the second is titled "Cuba" (1936). The second reference by León also details animal life and geology as well. Jan Thomas Roig's work titled "Diccionario Botanico" (1928) is an excellent generalist's guide to the flora of Cuba. A "Brief Overview of the Cuban Flora" was completed in 1940 by Mr. Carabia. The reconstruction of Cuba's original forests was a task that was very successfully done by German geographer Leo Waibel in his article titled "Place names to aid in the reconstruction of the original vegetation of Cuba" *Geographical Review* 33 (1943): 376–96. There is also Enrique del Risco Rodriguez' work on estimating Cuba's sixteenth-century vegetation in the *Nuevo atlas nacional de Cuba* and also for his book *Cuba: Su historia y caracteristicas* (1995). There are also the many references in each chapter that credit other scientists and geographers' work and helped supplement the work presented here. In order to give this work context it is important to showcase some of the notable work done before and any type of land cover or land use change begins with exploration and determination of what was where, and when.

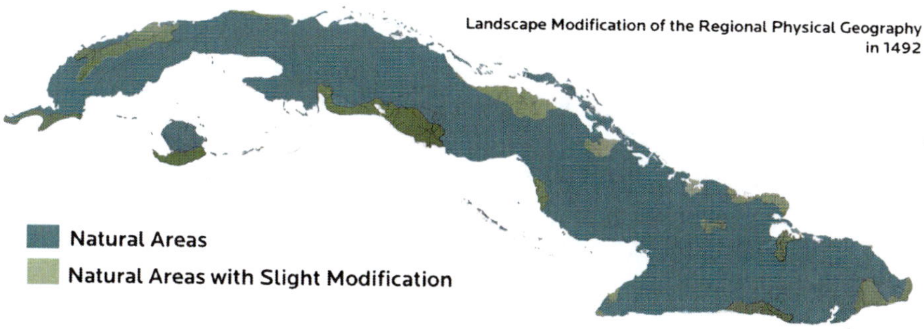

Fig. 1.2 An estimation of the first observed natural landscape extent in Cuba, 1492

Clearly from the prior discussion the planting and processing of sugarcane devastated the Cuban landscape more than any other factor. Combined with the increasing demand for tobacco and cattle grazing, these three factors were and still are reasons for concern regarding the health and long term outlook for habitats and the varied ecosystems that still exist throughout the island.

There are many ways to document land cover and land use change through time. Census data, compiled agricultural reports by the government, farmers' reporting crop yield per year, aerial photography taken every year over the same regions, these are all ways to document change. However, one of the most effective methods to document change is through a map. A map can be thought of as a snapshot in time. The best available data representing what is on the ground at a given moment. Enough of these types of maps over the same area spanning a long period of time can give an observer a sense of how the land cover and use has changed in an obvious, visual format. It is this format that is presented below to help compile the information that has been presented thus far.

The landscape of Cuba is quite complex and varied, this fact has been illustrated by its land use and land cover history which has evolved steadily since the arrival of the Spaniards in 1492. The original landscape of Cuba was much more homogeneous then that it is today as shown below (Fig. 1.2).

The following map summarizes the complex land use and land cover types that Cuba has become. The following map is generated utilizing ground data, census crop data, some aerial and satellite image acquisition and field data. It was generated using the online GIS data tools from Mapoteca Digital via Redcien (RedCien – Mapoteco Digital 1989).

It is always useful to drill down to more specific layers of information to the elements that characterize an area more specifically. So, for example while Fig. 1.3 shows us general land cover and land use classes the following map generated from the spatial data collection of the Nature Conservancy shows us more detailed categorizations of land cover in Fig. 1.4.

Finally, a more generalized map was created using Landsat Thematic Mapper data (described in later chapters) which is an orbital imaging platform that collects

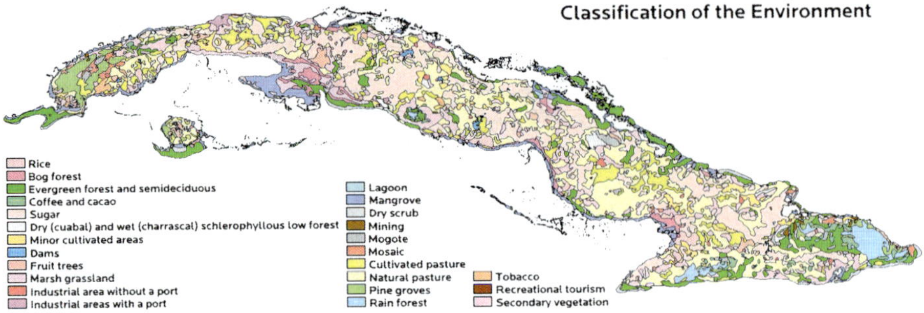

Fig. 1.3 GIS map showing many of the current land use and land cover categories

Fig. 1.4 This map shows more specific land cover classes as characterized by the nature conservancy (Map was generated by author using GIS data from TNC's spatial data archive here: http://maps.tnc.org/)

spectral information from things on the ground such as different kinds of plants, soils, urban areas, et cetera. There are excellent reasons to utilize satellite imagery such as that shown below in Fig. 1.5 for analyzing different types of land use/cover in an area. The first is there are many orbital systems that have been covering the same geographic areas for several decades. For instance Landsat acquires total global coverage every 15–17 days, acquiring imagery since the early 1970s, this means that it is possible to conduct time series analyses or simply have access to certain times and dates of interest. This was quite difficult in this case, over the Cuban archipelago, because in tropical areas there are many clouds much of the year.

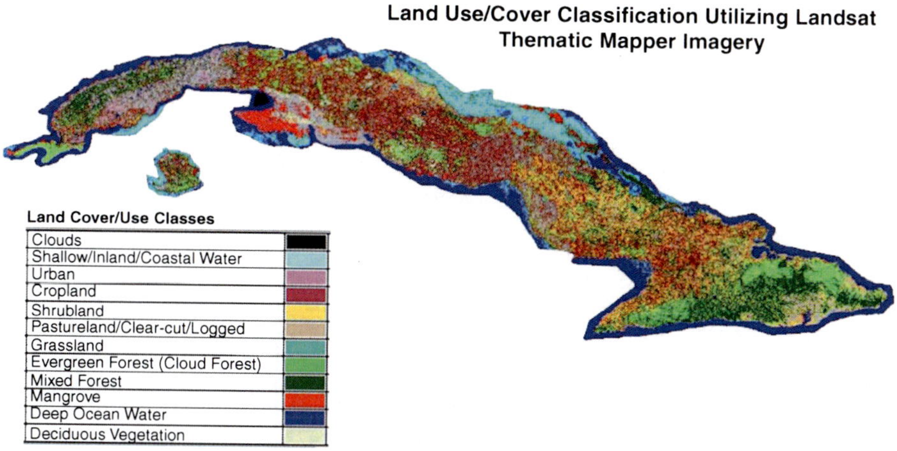

Fig. 1.5 Landsat classification of Cuba's land use and land cover utilizing mosaiced imagery from 1995 to 2001

Thus, because Landsat cannot acquire data through clouds, in depth time series over that entire 30-year period was not possible in minute detail using that satellite system. However, there is enough supplementary data that will show, in subsequent chapters, how it is possible to examine the changing features of Cuba's landscape and anthropogenic impacts.

In conclusion Cuba has changed significantly from its original landscape in the 1400s to the varied and much reduced natural landscape features that exist today.

References

Adams CK (1892) Christopher Columbus, his life and work. Cambridge University Press, New York

Bennett HH (1928) Some geographic aspects of Cuban soils. Geogr Rev 18:62–82
Stable URL: http://www.jstor.org/stable/208763

Consculluela J (1946) Prehistoric cultures of Cuba. Soc Am Archaeol 12:10–18. Stable URL: http://www.jstor.org/stable/275809

Crooker R, Pavlovic Z (2010) Cuba, 2nd edn. Chelsea House, New York

Day M (1978) Morphology and distribution of residual limestone hills (mogots) in the karst of Northern Puerto Rico. Geol Soc Am Bull 89:426–432. doi:10.1130/0016-7606(1978) 89<426:MADORL>2.0.CO;2

Díaz-Briquets S, Pérez-López J (2000) Conquering nature, the environmental legacy of socialism in Cuba. University of Pittsburgh Press, Pittsburgh

Domech R, Glean M (2001) Cuban nature. Publicaciones en languas extranjeras, Habana Cuba

Easterbrook D (1999) Surface processes and landforms, 2nd edn. Prentice Hall, Upper Saddle River
Encyclopædia Britannica. Encyclopædia Britannica Online (2011) "Casa de Contratación". Web. 08 Apr 2011. http://www.britannica.com/EBchecked/topic/135411/Casa-de-Contratacion
Fairbridge RW (ed.) (1968) Encyclopedia of geomorphology. Encyclopedia of Earth Sciences Series, vol. 3. Reinhold Book Corporation, New York, NY
Fewkes W (1904) Prehistoric culture of Cuba. Am Anthropol 6:585–598
Grupo Cubano de Investigaciones Económicas (1963) Un studio sobre Cuba. University of Miami Press, Coral Gables
Humboldt A (1856) The island of Cuba (English trans: Thrasher JS). Derby & Jackson, New York
Johnson WF (1920) The history of Cuba, vol 5. B.F. Buck & Company, Inc., New York
Keegan W (1995) Modeling dispersal in the prehistoric West Indies. World Archaeol 26:400–420. Stable URL: http://www.jstor.org/stable/124924
Luzon JL (1987) Economia, poblacion y territorio en Cuba (1899–1983). Ediciones Cultura Hispanica del Instituto de Cooperacion Iberoamericana, Madrid
Marrero L (1950) Geografía de Cuba. La Moderna Poesía, La Habana
Marrero L (1975) Cuba: Economía y Sociedad. El Siglo XVII (II), Vol 4. Editorial Playor, Madrid
Monzote R (2008) From rainforest to cane field in Cuba. University of North Carolina Press, Chapel Hill
Ortiz F (1947) Cuban counterpoint, tobacco and sugar. Duke University Press, Durham
RedCien – Mapoteco Digital (1989) Nuevo Atlas Nacional de Cuba. http://mapas.redciencia.cu:8081/mapoteca/. Accessed 2010–2011
Reed G (1992) On the razor's edge: deforestation in Cuba. Cuba Update 1–2:36–38
Rodríguez E (1999) Cuban forests their history and characteristics. Instituto Cubano del Libro, Havana
Schiettecatte W et al (2008) Influence on soil erosion risk in the Cuyaguateje watershed (Cuba). Catena 74:1–12
Seifriz W (1943) The plant life of Cuba. Ecol Monogr 13:375–426
Steinburg P (2001) The social construction of the ocean. Cambridge University Press, New York
Stevenson EL (1909) Early Spanish cartography of the new world, with special reference to the Wolfenbuttel-Spanish map and the work of Diego Ribero. The Davis Press, Worcester
Suchlicki J (1974) Cuba: from Columbus to Castro. Charles Scribner's Sons, New York
Thomas H (1998) Cuba or the pursuit of freedom. Da Capo Press, New York
Whitbeck R (1922) Geographical relations in the development of Cuban agriculture. Geogr Rev 12:223–240

Chapter 2
Linking Causal Factors to Areas of Highest Change

Abstract This chapter focuses on several key incidents in time that helped shape the landscape of Cuba into law, policy and future landscape characterization. Here reasons for change from the time of the indigenous Indians to current impacts are briefly described and discussed. This chapter begins the dialogue pertaining to linkages between causal factors and areas that exhibited the highest levels of change over time on the Island of Cuba.

J. Gebelein, *A Geographic Perspective of Cuban Landscapes*, Landscape Series 15,
DOI 10.1007/978-94-007-2406-8_2, © Springer Science+Business Media B.V. 2012

Keywords Governor Diego Velázquez • Royal Forest Reserves • Haiti's slave revolt • Reforestation project • Special Period • Torricelli bill "Cuban Democracy Act" • 1992 Rio Summit • Green agriculture • Integrated pest management (IPM) • Development, Sustainability and Equity Meeting • Cauto River Basin • Sustainable Energy Development Meeting

2.1 Introduction

Chapter 1 focused on where changed occurred on the island of Cuba from the time of Columbus onward. This chapter is focused on the causes of those changes through time. Here the focus is on several landmark incidents, laws and policies, or situations which sparked a great deal of change that impacted the landscape for years in its wake. There were many reasons for landscape transformation, and the modification of land depended on several key factors throughout history. The first controlling factor is the ruling population which is directly linked to the population of the ruling people. For example, as illustrated and described in Chap. 1, the three groups of Indian tribes that existed in the late 1400s appeared to have a very low impact on Cuba's environment. It was largely due to the fact that the Ciboneyes and the Guanahatabeyes did not rely much, if at all, on agriculture and their numbers were quite low. They were both primarily hunting and gathering cultures. The Tainos were a much larger group and did alter the landscape utilizing various agricultural practices such as slash and burn agriculture. Their numbers were also much higher than the Ciboneyes or Guanahatabeyes. However, despite the existence all three groups, it is important to note that the forests and natural landscapes of Cuba were observed to be over 90% intact in the late 1400s. Thus clearly the natural resources of the island were sufficient for their populations and sustainable for the various ecosystems in Cuba in the 1400s.

2.2 Reasons for Significant Landscape Changes

With the introduction of the Europeans beginning with Columbus and then the period of Spanish rule into the 1500 and 1600s came changing demands on Cuba's resources that were shaped into law and policy by the Spanish crown. One of the events that shaped the landscape of Cuba was the introduction of African slaves beginning in the late 1500s. This was a pivotal point in Cuba's history. With this practice the Spanish then had a steady supply of labour for the highest yield possible from their lands. The steady increase in bringing in African slaves from that point became the foundation for the evolution from sedentary farming to industrial agriculture (Monzote 2008). Before discussing the reasons for sugar's impacts on the environment, it is important to note that although sugar was *introduced* as a crop very early on by Governor Diego Velázquez in the early period of Spanish colonization, it was not until 1590 that it began to be planted for more industrial purposes. And even after this larger effort began to cultivate sugar, its growth was quite slow initially (Rodríguez 1999). Before that time Cuba's land was primarily serving the needs of cattle ranchers and livestock farms, and animal husbandry was one the of two major activities changing the landscape. The second major factor was logging wood. In the late 1500s there was a wood crisis in Spain. The most impacted areas were those near major ports in which forests were logged of timber for ship building, precious woods for royal construction projects and local fuel requirements. However, even with these early impacts the most significant influence on the landscape with regard to permanent environmental alteration is sugar.

Sugar was grown in small amounts on for a very long time before its explosion and subsequent replacement of many natural habitats over large areas in Cuba. In 1595 two major incidents occurred that marked the beginning of the sugar industry in Cuba. First was the authorization to introduce black slaves into the Indies. The permission was granted by Felipe II to Gómez Reyell of Portugal. Later that same year, Felipe II also decreed that Cuba's mill owners should have the same rights granted to owners in Hispaniola, specifically, that their tools, machines, animals, lands or slaves could *not* be appropriated because of debt. In addition to these two important events, in 1602 the acquisition of a royal loan to stimulate sugar production was granted to Havana's first plantation owners (Monzote 2008). During sugar's evolution the industry's growth in Cuba was fundamentally dependent on the Royal Forest Reserves. These reserves had control over much of the island's forested lands for the purpose of ship building to strengthen the empire. Sugarcane production had little power over the ship building industry and until Haiti's revolt in the 1790s. The slave revolt in Haiti spurred sugarcane growth in Cuba. In fact we can link this revolt and sugar's subsequent huge production increase in Cuba as the beginning of the extensive obliteration of Cuba's natural habitats, especially forests. This period of time where different types of forests were replaced with sugarcane crops lasted just over 100 years from approximately 1800 until the 1920s. During this time forests were logged or (usually) slashed and burned with the sole purpose of increasing the land devoted to sugarcane (Díaz-Briquets and Pérez-López 2000).

With regard to forest change in 1959, Castro began a reforestation project shortly after assuming power. As noted in Chap. 1, this effort was successful and that success was due to several factors. The first is that the Cuban government officially reserved 67.6% of national forests and placed them under the classification of protected areas. They also set aside 32.4% of the national forests for timber production (COMARNA 1992). The result of these actions was that the forests were now managed in two capacities, one for preservation and one for consumption. More specifically, the harvesting rates were reduced from 150,000 cubic meters to 45,000 cubic meters per year. The net addition to the forested woodland areas was almost 14,000 ha annually between 1959 and 1992. However, there were obstacles to this reforestation such as the sugar harvest in 1970 that resulted in the number of trees being felled outnumbering the trees being replanted (Rodríguez 1999). According to the government's reforestation program statistics, between 1960 and 1990, 2.5 billion trees were planted. COMARNA (1992) declared that for every hectare of forest that was logged, 16.9 ha of juvenile trees were planted. This translates into for every 6,500 ha consumed, 110,000 ha were planted (Reed 1992). Assuming that this information is accurate, this means that in the 1990s the annual reforestation rates were very similar to the deforestation rates in the 1910s and 1920s. An official report by the Economic Commission for Latin America and the Caribbean (CEPAL) asserts that Cuban forest cover has sustained this gradual increase attaining a total area of 2.7 million hectares (or 24.5%) in 1995. According to Diáz-Briquets and Pérez-López (2000) this is not accurate and is an interesting case of deliberate reclassification of other land cover classes into the forest land cover class to boost these numbers. The reclassification of formerly cultivated lands, subsequently abandoned, during the special period due to the lack of imported agricultural goods. Therefore there is an academic debate whether these reports of current forested lands are actually correct and covers about 18–20% of Cuba's natural areas, versus some experts now believe that the percent of forest is well below this between 8% and 10% of Cuba's land cover (Diáz-Briquets and Pérez-López 2000; Domech and Glean 2001; Rodríguez 1999; Thomas 1998).

The forests that were most impacted over time were those that were easily accessible. This meant that the forested areas in lowlands, in the centre and western Cuba, and those in the lower elevations in the east, or those that were located near major rivers (for easier transport) were the most impacted. In regions such as the Sierra Maestra for example, which are steep and difficult to access, include some of the more pristine areas of natural forest left in Cuba. Clearly the other issue in these flatlands was the establishment and boom of sugarcane. More forest was cleared for the use of sugarcane than any other crop.

As discussed previously, agricultural practices and the policies that govern them have become one of the most significant categories of impact spanning the entire history of Cuba. Research conducted by both Cuban and international scholars has provided data and analysis that has narrowed the focus of recent agricultural transformations into several periods of time; we will begin with the Special Period because of its level of influence on Cuban landscape alteration and ultimately, its future.

The Special Period in Cuba (*período especial*) lasted in Cuba from approximately 1990–2000. In short the events that preceded the Special Period were largely the fall of the Soviet Union and the strengthening of the United States blockade. What truly drove Cuba into crisis were the suspension of economic relationships with the former Soviet Union and the loss of trading partners. Approximately 85% of Cuban foreign trade had been with the socialist bloc of Mutual Economic Assistance (CMEA) and the majority of those transactions were usually completed in nonconvertible currency. This severely impacted the Cuban economy because commercial trade dropped by greater than 90%. In 1989 trade was $8.7 billion, this dropped to $4.5 billion in 1991 followed by only $750 million in 1993. Additionally, the Soviet oil imports dropped by 90% from 13 million tons in 1989 to 1.8 million tons in 1992. All types of consumer goods, basic foodstuffs, and grains fell significantly and the importation of spare parts and raw materials vital to many different Cuban industries stopped completely. Agriculture was severely impacted due to a decline in fertilizer imports by 80% in a very short period of time. This decline measured from 1.3 million tons dropping to 25,000 tons and animal feed supplies declined by 70% from 1.6 million tons to 450,000 tons in that same short period. This series of events led to extreme scarcity of goods and fuel shortages affected the closing of factories and industrial plants.

Harvesting equipment, tractors and trucks were replaced by farm animals because of the lack of spare parts and access to fuel. And agricultural production was greatly reduced because of the shortage of herbicides and fertilizers (Pérez 1995).

The Cuban economy and its people had benefited from the relationship with the former Soviet Union, however, this relationship was truly a dependency and when trade relations with the soviet bloc evaporated, Cuba became stranded in the global marketplace. The United States further isolated Cuba by issuing an executive order in 1992 that prohibited ships trading with Cuba to enter U.S. ports. This was an effective technique to reduce the number of vessels (and countries/companies) wishing to conduct trade with Cuba (Thomas 1998). In that same year the U.S. Congress enacted the Torricelli bill or "Cuba Democracy Act" as it is sometimes known. This stipulated that subsidiaries of U.S. companies that function in third countries were forbidden from trading with or investing in Cuba. In addition this law sanctioned the U.S. President to withhold debt relief, economic assistance and all free trade agreements with countries that give aid to Cuba (Pérez 1995).

Clearly this was an extremely difficult period for all Cuban people and the magnitude of these events led to distinct changes in the landscape during this time frame from 1990 to 2000. These changes were linked to different environmental issues that occurred before and after the Special Period. In 1990, when the Special Period took effect in Cuba, the environmental situation was quite serious. Deforestation was a continued challenge in spite of efforts by Castro to reforest areas of interest through a formal program. In fact, one of the most significant causes of species endangerment and extinction in Cuba was deforestation and habitat fragmentation (Tuxill and Bright 1998). When the Special Period began; there was a decline in numerous environmentally damaging activities by both necessity and by choice. Prior to the Special Period, there were elevated risks to the environment posed by industrial wastewater and sludge, land-based degradation, freshwater quality issues, pesticides and surface water stressors. The bays were quite polluted from these types of activities. This marine pollution reduced the marine and coastal habitats for juvenile fish species, rookeries for juvenile birds, and damaged mangrove habitats. In addition was the impact from these activities on human health. And there were several areas, such as in Holguin and Moa that were environmentally destroyed following the installation and operational activities of strip mining.[1] Following the Special Period, pesticides, sludge and industrial wastewater were no longer a prominent set of concerns for the environment; however marine coastal deterioration and municipal wastewater

[1] I can personally attest to the severe impact of Moa's mine which, in 2006 identified its presence more than 2 miles away by large stands of dead trees, no birds whatsoever in the area, standing water that reeked of chemicals and an odd crust that was on most exposed soil surfaces as we approached the mine.

became higher-level concerns than in the past. There was a decrease in dumping of waste into bays and rivers, a reduction in gaseous emission into the atmosphere, and diminishing overexploitation of water resources (CITMA 1992).

The major economic directions taken by the Cuban government at that time were to implement cutbacks in state spending, legalization of foreign investment and dollar possession, promotion of tourism and beginning programs of self-sufficiency regarding food (Dello Buono 1995). Certainly the Special Period has also brought some negative impacts on the environment as well. For instance between 1992 and 1995 deforestation *increased* in very environmentally sensitive areas in the Zapata Swamp and nearby mangrove forests. In terms of negative marine impacts due to the Special Period, since there was such a decrease in available consumer goods, there was a large spike in the statistics regarding fishing in protected areas, and sale of exotic aquarium species (CITMA 1992). Maal-Bared (2006) conducted a comparative environmental risk assessment which analysed and ultimately ranked marine and terrestrial environmental problems before and after the Special Period. Interestingly, terrestrial degradation was the top environmental problem before *and* after the Special Period. In his analysis the author defines terrestrial degradation as several key elements including deforestation, dam construction, mining, tourism, military operations and agriculture.

The government of Cuba has been actively engaged in international collaborative efforts to conserve its environment by supporting its scientists, researchers and natural resource managers in land and coastal resource conservation. For example, the Cuban government was actively engaged in the 1992 Rio Summit. It was after this meeting that more environmentally-related institutions were created as well as new and improved policies, laws and programs influencing environmental impact assessments, management of protected areas, issues with forest preservation and logging and environmental education (Lane 2000). The link between conservation of natural resources and sustainable development became a high priority for the government. For example, Castro turned his government's attention and energy to organic farming and more reliance on renewable resources. Even though this kind of research began in the early 1980s, it received intense scrutiny and encouragement during the Special Period. Organic techniques of farming are usually better for soils in terms of nutrient replenishment and lack of accumulation of pesticides in the groundwater, rivers and streams. So clearly this type of agricultural development has a positive impact on water quality, soil nutrient stability, and decreased soil salinization, for example. Two types of Cuban "green" farming methods will be shown here. The first is called integrated pest management (IPM) and is the integrated and multifaceted use of several pest, weed control and plant disease approaches for the purpose of reducing dependence on chemical pesticides. Cuban scientists[2] have developed

[2] There are many institutes who support plant and crop protection research. The Ministry of Agriculture houses the National Service of Plant Protection, the Central Research Laboratory and the Institute of Plant Protection, as well as 14 regional labs and 60 plant protection stations all over the island (Rosset and Benjamin 1994).

natural pesticides that are extracts from plants, particularly the neem tree. At the end of 1991 during the Special Period, approximately 56% of Cuban agriculture was treated with organic biological controls. The second example is rotating crops for the purpose of out-competing the weed community. An example of this is planting corn or sorghum, then once that has been harvested planting potato, harvest, and then back to corn or sorghum once again. Doing this, in combination with application of natural herbicides, is recommended to control feverfew or Mugwort (*Parthenium hysterophorus*) and additional Dicotyledonous annual weeds (Rosset and Benjamin 1994). Clearly the Special Period had a large impact on the environmental landscape of Cuba, some of it negative and other aspects positive. However it is clear that the repercussions of the Special Period can still be seen today pertaining to organic farming techniques, and the shift in pollution sources and vulnerabilities on the terrestrial as well as marine lands.

Another recent meeting made an impact on the Cuban governance of the environment. It was the "Development, Sustainability and Equity" IPCC Expert Meeting held in Havana in February 2000. Some of the more notable results of this meeting were that several countries like Cuba laid the groundwork for the introduction of measures that were decrease energy intensity of GDP and this reduce emission. This was a major milestone for Cuba particularly in the industrial sector of the economy (IPCC 2000).

Most of the environmental examples in this manuscript have received global notice; however it is important to provide at least one localized example of a recent pollution milestone that marks great change in an area. This refers to the Cauto River Basin. The Cauto River Basin exemplifies many of the agricultural problems that are described in this book. This is considered one of the most environmentally compromised areas in Cuba. The Cauto Basin is approximately 8% of Cuba's territory and makes up 9.3% of its agricultural land. Roughly 10% of the sugar and 25% of the rice in Cuba is cultivated here (Díaz-Briquets and Pérez-López 2000). The natural runoff has been decreased by 60% in only a few decades, there are almost no forest areas anymore, salt water intrusions have ruined most of the groundwater reservoirs and one-third of this region is subject to severe erosion (Portela 1997). Additionally, Portela (1997) found that the water there is critically polluted by effluent from 652 sources including urban sewage, cattle farms and industrial waste. Many sources believe that the reason for the current state of the Cauto Basin is because of bad management practices when developing hydraulic projects, a circumstance that was only made worse by deforestation. This situation led to a 2 year study in 1997 to resolve the extent of the damage and decide how to begin reversing or at least stop this environmental deterioration (Riera 1998). This situation received much attention nationwide and I believe was a warning to the Cuban government that this is what their island could become, if this kind of management became commonplace.

Even more recent was Cuba's participation in the "Sustainable Energy Development" Meeting in Vienna on assessing Cuba's mechanisms for incorporating the concepts of sustainable development into practical implementation strategies. The assessment was specifically directed at one of the most important sectors affecting economic and social development—energy. The major findings can be summarized by stating that Cuba has a high level of electrification (95.5%), urbanization, industrialization, and reforestation good access to sanitation services. Despite the severe predicament in the 1990s Cuba's refocusing on domestic fuel supplies instead of imports and fuel conservation for rural areas, was hopeful. Thus, Cuba has demonstrated that it has a high adaptive capability to overcome substantial socioeconomic situations caused by the collapse of trade with the Soviet Union. However the negative environmental influences due to the large emitters of pollutants from industrial processes are not getting better. In fact there are areas with very high concentrations of pollutants and deteriorating air quality, and impacting human health and environmental conditions overall (International Atomic Energy Agency 2008).

Cuban officials took this opportunity to begin working with members of this panel to cooperate in amending old strategies and creating new approaches to energy frameworks and structures for the future.

References

CITMA (Ministry of Science Environment and Technology) (1992) Referring to the negative repercussions of the blockade in the development and environment field. Cuba' Academia de Ciencias de Cuba, La Habana

COMARNA (1992) Informe nacional a la conferencia de Naciones Unidas sobre medio ambiente y desarrollo Brasil/1992. Resumen, La Habana

Dello Buono RA (1995) An introduction to Cuba in the special period. In: Dello Buono RA (ed) Facultad Latino Americana de Ciencias Sociales Programa Cuba. Carta Cuba: interdisciplinary reflections on development and society, La Habana

Díaz-Briquets S, Pérez-López J (2000) Conquering nature, the environmental legacy of socialism in Cuba. University of Pittsburgh Press, Pittsburgh

Domech R, Glean M (2001) Cuban nature. Publicaciones en languas extranjeras, La Habana

International Atomic Energy Agency (2008) Cuba: a country profile on sustainable energy development. International Atomic Energy Agency, Vienna

IPCC (2000) Second regional expert meeting on "Development, equity and sustainability" in havana, Cuba, 23–25 Feb

Lane P (2000) Green Cuba. Peace Rev 12(2):319–324

Maal-Bared R (2006) Comparing environmental issues in Cuba before and after the special period: balancing sustainable development and survival. Environ Int 32:349–358

Monzote R (2008) From rainforest to cane field in Cuba. University of North Carolina Press, Chapel Hill

Pérez L (1995) Cuba – between reform and revolution, 2nd edn. Oxford University Press, New York

Portela A (1997) Cuba assesses damage, finds it excessive, Cuba news 5 (7) July: 11. Postergada indefinidamente la terminacion de la planta de Juraguá

Reed G (1992) On the razor's edge: deforestation in Cuba. Cuba Update 1–2:36–38

Riera L (1998) Will its waters ever run clear again? Granma electronic edition 3 (March)

Rodríguez E (1999) Cuban forests their history and characteristics. Instituto Cubano del Libro, Havana

Rosset P, Benjamin M (eds) (1994) The greening of the revolution – Cuba's experiment with organic agriculture. Ocean Press, Totowa

Thomas H (1998) Cuba or the pursuit of freedom. Da Capo Press, New York

Tuxill J, Bright C (1998) Losing strands in the web of life. In: Starke L (ed) State of the world. Worldwatch Institute Publication, New York

Chapter 3
Governmental Organization and Control Over Environmental Policies

Abstract Cuba has a vast number of agencies controlling its interior lands, coastal and maritime areas. This chapter describes the framework of these agencies, institutes and how their existence is linked to catalysts such as the Earth Summit in Rio de Janeiro in 1992. There are also two maps in this chapter that illustrate the year protected areas were established as well as what general entity is in charge of each protected area. The chapter concludes with the Ministry responsible for human health issues. The reason for its inclusion here is the direct link between a healthy environment and a healthy population.

J. Gebelein, *A Geographic Perspective of Cuban Landscapes*, Landscape Series 15,
DOI 10.1007/978-94-007-2406-8_3,

Keywords Human health • Mining • Pollution • Ministry of environment • CITMA • Law of the Environment (Law 81) • Article 86 • The Cuban Ministry of Public Health (MINSAP) • Ministry of Informatics and Communications (MIC) • Ministry of Tourism (MINTUR) • Ministry of Economics and Planning • Ministry of Sugar • Ministry of Agriculture • Ministry of the Fishing Industry • Article 11 • Earth Summit

3.1 Introduction

Cuba's environment is complex and consists of many microcosms of weather, soils, plant and animal life. Due to this complexity the management of these many ecosystems has evolved in an equally multifarious manner. The purpose of this chapter is to help deconstruct some of that complexity. Part of this deconstruction includes explaining the fundamental laws and policies that have helped shape these agencies. Additionally there have been several major international meetings that have shaped international cooperation and in turn, the development and establishment of these ministries and agencies of protection. There are a myriad of control levels and active versus passive protection of terrestrial lands, coastal areas and waters, thus we begin with the specific focus on conservation management and protection of these areas and species within them.

3.2 A Specific Focus on Conservation Management, and Protection of (Flora/Fauna) Species

The Cuban government is divided into many categories which manage and control the natural resources, land and waters that belong to Cuba. It can be very confusing to separate out levels of structure and controls over different habitats, agriculture, conservation measures and protections for the flora, fauna, and geophysical aspects of Cuba. Hopefully the information provided here will make clearer the overall organization and structure regarding environmental control and policies.

Cuba has a biological diversity that is extraordinary. It is an extremely important habitat for thousands of species of birds, mammals, fish, vegetation communities and microclimates that accommodate many migratory species as well. In terms of biodiversity opportunities for scientific research as well as cultural heritage, Cuba is as critically important. It is interesting to also look at a graphic depiction of Cuba's consistent establishment of protected areas throughout the island as well as marine areas. The figure below (Fig. 3.1) clearly shows that, at least since 1900, there have been regular additions to the number of preserved areas on land and at sea. This is largely due to the realization of the Law of the Environment, acceptance of the goals for the National System of protected Areas (SNAP) as well as the establishment of the Cuban subsystem of Marine Protected Areas (SAMP).

Regarding control of all these protected areas it is quite clear from the following map (Fig. 3.2) that the governing body of these protected areas are almost exclusively

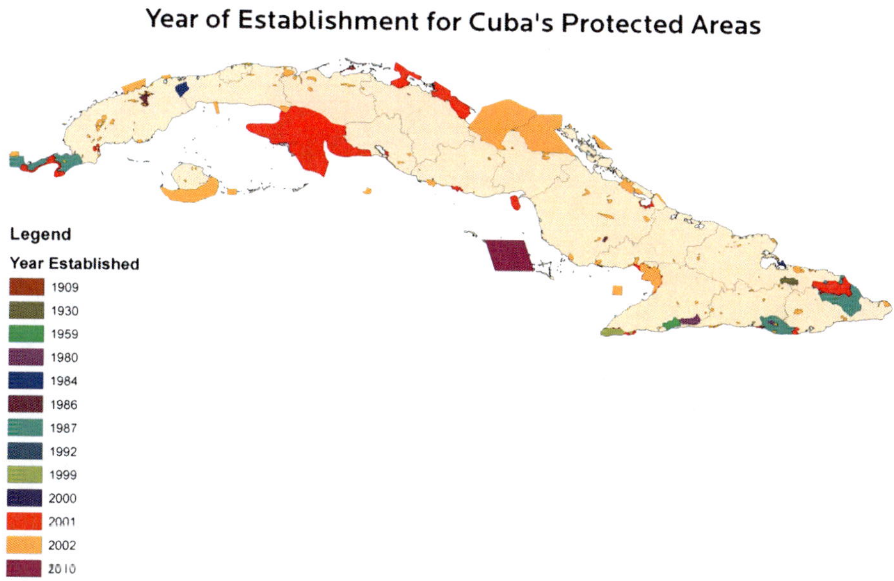

Fig. 3.1 GIS map illustrating year that protected area was established

Fig. 3.2 Entity in charge of protected area

government run entities. This is something that the Cuban government is attempting to change by involving more international, regional and local collaborative research groups. However at the time of this book, this map still represents the reality of controlling agents for most protected areas in Cuba.

Cuba's natural resources are quite impressive with almost 4,000 miles of coast on the main island and several thousand outlying keys around its periphery. The shorelines in particular have become a major attraction for international tourists. In fact, by 1996 agriculture was no longer Cuba's principal source of revenue; it was tourism (Crespo and Negrón-Díaz 1997). Before delving into international impacts however, it is useful to understand why the Cuban laws and policies look like they do today.

One of the main purposes of the Cuban Revolution was to remove systems and associated laws that were considered unjust and anti-social. One of the first major targets were the issues associated with the practice of law and private property. Bearing this in mind the Cuban government began to move towards reorganization, redistribution of lands and a planned economy, and this necessitated a different kind of state (Houck 2000). It is from this mindset that many of the laws, policies and governing bodies were derived. There is overlap and some shared responsibilities in several of these organizations in terms of terrestrial and marine issues.

In 1976 the National Commission for Environmental Protection (Comisión Nacional para la Protección del Medio Ambiente y los Recursos Naturales: COMARNA was created, then subsequently replaced by the Ministry of Science, Technology and Environment (Ministerio de Ciencia, Tecnología y Medio Ambiente: CITMA) in 1994. In 1997, the National Environmental Strategy was approved and served as a baseline in developing and implementing a succession of environmental laws and regulations putting into practice a new paradigm of natural resource protection (Whittle and Santos 2006; Cruz and McLaughlin 2008). Thus, knowing this context we can ask the question: "What laws led to the current environmental protections we see today, what are the key government organizations, and what is their control over environmental policies?"

The Earth Summit in Rio de Janeiro in 1992 was the basis for most current environmental law in Cuba. Sustainable development was quickly adopted by the government leaders who attended and later that same year the Cuba constitution was amended to fortify proper protections for the environment. In 1994, the Ministry of Science, Technology and Environment (CITMA) was created as a first cabinet level ministry. The first task of CITMA was to conduct an assessment of the current environmental conditions and document extent and damage of natural resources. Based on this the National Environmental Strategy was created as a basis for an inclusive environmental set of laws and policies (Whittle and Lindeman 2004). CITMA's responsibilities are outlined in the Law of the Environment (Law 81) and Article 11 makes CITMA the lead entity for all environments in Cuba. It states:

> The Ministry of Science, Technology and the Environment is the governmental agency of the Central Administration of the State in charge of proposing environmental policy and guiding its execution through the coordination and control of the nation's environmental management, promoting its coherent integration in order to contribute to sustainable development.

CITMA has, for example, has the power to veto projects pertaining to the construction of new hotels and other developments that it determines are not environmentally rigorous (Whittle et al. 2003). CITMA has been responsible for a number of key laws (summarized in Chap. 4), and a number of important environmental bodies, which will be summarized here.

CITMA is comprised of three central offices: (1) the Environmental Agency, (2) Office of Regulation and (3) the Environmental Directorate. The Environmental Directorate is in charge of creating and overseeing all environmental regulations, policies and laws. The Office of Regulation manages compliance with all environmental rules. CICA resides within the Office of Regulation. When an investor wants to build a new hotel, for example, he must submit an application to the Center for Inspection and Environmental Control (CICA) within CITMA. CICA decides whether to approve this application and grant a license to build *or* submit the application to the Environmental Impact Assessment (EIA) which assesses any project to identify possible environmental impacts. The Environmental Agency supervises numerous educational, research and scientific institutions. The Environmental Agency manages the National Center for Protected Areas, whose chief responsibility is to manage terrestrial and marine parks and additional protected areas. CITMA also operates a National Coastal Group composed of 14 ministries and additional administrative bodies (Whittle et al. 2003).

The Law of the Environment served as a cornerstone for founding the guiding principles for the improvement and conservation of non-living and living resources within the coastal zone. This was then augmented by the Law of the Coastal Zone (Law 212) established in 2000, which specifically protected the coastal zones.

Based on these laws, scientific workshops and international collaborations, the "Coastal Zone Management" was further designed to function as judicial tool for sustainable use and conservation of all coastal areas in Cuba. The National Coastal Group, Institute of Oceanography, Ministry of Fisheries, and the Center for Engineering and Management of Bays have begun to supply advice and significant influence for policy action plans that enact integrated administration of these areas for the entire island (Lindeman et al. 2003).

For the reader's reference some of the most relevant organizations are listed and described here.

3.2.1 *Ministry of the Fishing Industry*

The Ministry of the Fishing Industry is the organization responsible for managing, executing and examining government and state policy associated with extraction, conservation, exploration, cultivation, processing and marketing of fishing resources and the related merchant fleet. In 1995 the Ministry of Fishing Industry instituted a series of organizational changes which were necessary to improve financial management, commercial activity and efficiency in production. That new structure and in depth description can be found in Adams et al. (2000).

3.2.2 *Ministry of Agriculture*

The Ministry of Agriculture is the entity accountable for managing and guiding policies associated with the preservation, and cultivation and improvement of forestry and agricultural lands pertaining to health of livestock and associated medicines, any use or management practice associated with of forest resources, flora and fauna within an agricultural or forest area, overall plant health and vigor, improving irrigation techniques and programs, encouraging farmers to cultivate crops other than forestry, sugarcane or grazing lands, oversee the growing and marketing of poultry, citrus, rice, tobacco, and other crops.

3.2.3 Ministry of Sugar

Since sugar has had such a dramatic impact on the landscape of Cuba it has been assigned its own governing entity that only focuses on sugarcane and its production. The Ministry of Sugar is the organization accountable for regulating, implementing and examining state and government policy related to sugarcane cultivation and the processing of sugarcane and its derivatives, with the goal of achieving sustainable development of production, in order to meet domestic and export demands.

3.2.4 Ministry of Economics and Planning (Previously the Ministry of Planning)

Land use planning is managed primarily by the Ministry of Planning and is done at the municipal, provincial and national levels. They are responsible for generating guidelines, policies and priorities for land development and use and should be in harmony with financial plans for the same areas. Part of Law 81 encourages planners to work with environmental officials. This institute has very specific rules and regulations for tourism in particular. Whittle et al. (2003) describe this ministry in exhaustive detail.

3.2.5 Ministry of Tourism

The Ministry of Tourism (MINTUR) was created in 1994 under the jurisdiction of Decree-Law 147 is the body responsible improving marketing, diversifying tourism options to become more competitive, increasing hotel rooms on Cuba, updating computers and communications networks to adhere to international standards, bring more foreign investment to Cuba's tourism sector. MINTUR has a two-tier hierarchy. The first is administrative with the minister and vice ministers and the second is individuals responsible for legal issues, commercial development, advertising and marketing, economic research, investments, and international relations (Ministry of Foreign Affairs, MINTUR 2003). In summary MINTUR is accountable for guiding, assessing, monitoring and executing, state and government policy associated with tourism.

3.2.6 Ministry of Informatics and Communications (MIC)

This is the governing body of computer science and communications in Cuba. On a side note here, according to an article "The Internet in Cuba" (Press 2011), China is in negotiations with Cuba to upgrade an undersea cable for internet. If this occurs it will transform the Cuban people's capacity to access information. However, the computer infrastructure is currently insufficient in Cuba to make this a real issue (Press 2011).

There are many institutes which also fall under CITMA's management umbrella which include the Institute of Meteorology Instituto de Meteorologia (INSMET), the Institute of Geophysics and Astronomy (Instituto de Geofísica y Astronomía), the Institute of Scientific and Technological Information (Instituto de Información Científica y Tecnológica (IDICT)), the Institute for Ecology and Systematics (Instituto de Ecología y Sistemática (IES)), and the Institute of Meteorology (Instituto de Meteorología de la República de Cuba).

There are also many international organizations that work extensively with the Cuban ministries, institutes and research centres such as the Environmental Defense Fund (EDF) which had been instrumental in collaborative work on marine and terrestrial assessments. The United Nations Environmental Program in Cuba, The Nature Conservancy, The Harte Institute, and the United Nations Food and Agriculture (FAO) organization now has an office in Cuba as well. All of these organizations provide collaborative assistance to Cuba regarding the environment and humanitarian issues as well. It is useful to also address the human health links to environmental policies since many of the same environmental issues discussed directly impact an individual's health.

3.3 Human Health Links to Environmental Policies

The Cuban Ministry of Public Health (MINSAP) is the primary organization in the Cuban government that is responsible for managing, administering, updating and controlling the application of the State and Government's public health regarding medical sciences and research and development pertaining to the Pharmaceutical Industry. Cuba's health industry has changed quite dramatically. From 1989 to 1993 the availability of medical equipment and supplies dropped by 70% and Cuban's daily caloric intake also dropped sharply by 33% overall. The disappearance of the Soviet's trade relations with Cuba created fuel shortages and thus water pumping stations stopped working regularly and had to be rationed. Ultimately food was rationed and very scarce. This had a striking impact on Cubans and while things have improved since then, it is because the Cuban government has made an effort to reinvest in their medical programs and augment classic medical care with homeopathic remedies.

The above is a brief summary of the government link to human health but there is also the environmental potentials that could impact Cuba environmental policies such as climate change. Climate change can have serious effects on population health as well as many characteristics of social and ecological systems (Bultó et al. 2006). For example climate could potentially generate certain conditions that encourage the development of disease-causing microorganisms (McMichael and

Kovats 1999). Bultó et al. (2006) conducted a study that illustrated that some disease not previously thought to be sensitive to climate change actually are including: chicken pox, bacterial and viral meningitis and others.

The importance of recognizing the linkages between the environment and human health is great. The Cuban government has begun to do this in the past several decades with the introduction of the laws and policies of the environment in conjunction with the organizational structures that govern the island's flora and fauna, and its conservation. This concept is shown in the Law of the Environment (Law 81) in Chap. 2, Article 86 where it states:

"**Article 86**. It is the responsibility of the Ministry of Science, Technology and Environment, in coordination with the Ministry of Agriculture and other responsible agencies and bodies, to issue provisions related to the importation and introduction into the environment of new species or species submitted to special regulations, to which ends the following principles will be taken into account:

1. the possible reactions of the species in the environment in which they will be introduced;
2. the possible reactions of the recipient environment and of native species to the species that will be introduced;
3. the risk that potentially dangerous genotypes may arise;
4. the possible introduction of exotic and epizootic diseases that may affect plants or animals;
5. the risk to human health;
6. other principles of special interest for the protection of the environment." Houck (2000)

There are several major sources of pollution that are linked with human and environmental health issues in Cuba. Cuban sugar mills are well known polluters of streams, reservoirs, wetlands and coastal areas with liquid waste effluents such as bagasse composts, fermented liquid wastes, molasses, and filter mud. This kind of waste especially impacts overall water quality and marine resources. Mining is another industry that can have negative impacts on humans and their environment. Mined areas, particularly nickel, are denuded of any vegetation and are not capable of absorbing rainwater which allows remaining soil to be washed away in rainstorms causing further erosion to the area. Studies have shown that when exposed to moisture (like rain) the mined areas exude acids that are then carried by the rainwater to rivers and streams in close proximity (Díaz-Briquets and Pérez-López 2000).

Cuba also has several chemical plants which, like the "Patricio Lumumba" chemical plant in Matahambre, discharge large amounts of hydrogen sulfide (including heavy metals) into the atmosphere and industrial effluent onto the marine shelf. In this case the nearby towns of La Sabana and Santa Lucia have been impacted by these air pollutants and the inhabitants exhibit high incidences of pulmonary diseases and asthma (Oro 1992).

These types of connections exist and it is imperative that these obvious associations between pollution, human health and policy be strengthened so Cuba can move into a healthier, integrated and more comprehensive system of management.

References

Adams C, Vega PS, Ministerio de la Industria Pesquera, Alvarez AG (2000) An overview of the Cuban commercial fishing industry and recent changes in management structure and objectives, Centro de Estudios de la Economia Cubana, Universidad de la Habana IIFET proceedings

Bultó PLO, Rodríguez AP, Valencia AR, Vega NL, Gonzalez MD, Carrera AP (2006) Assessment of human health vulnerability to climate variability and change in Cuba. Environ Health Persp 114:1942–1949. doi:10.1289/ehp. 8434

Crespo N, Negrón-Díaz S (1997) Cuban tourism in 2007: economic impact. In: Association for the Study of the Cuban Economy, Cuba in transition 150:152

Cruz I, McLaughlin RJ (2008) Contrasting marine policies in the United States, Mexico, Cuba and the European union: searching for an integrated strategy for the Gulf of Mexico region. Ocean Coastal Manage 51:826–838

Díaz-Briquets S, Pérez-López J (2000) Conquering nature, the environmental legacy of socialism in Cuba. University of Pittsburgh Press, Pittsburgh

Houck O (2000) Environmental law in Cuba. J Land Use Environ Law 16:1

Lindeman KC, Tripp JTB, Whittle DJ, Moulaert-Quiros A, Stewart E (2003) Sustainable coastal tourism in Cuba: roles of environmental assessments, certification programs and protection fees. Tulane Environ Law J 16(Special issue):591–618

McMichael AJ, Kovats S (1999) El tiempo el clima y la salud [in Spanish]. Bol Organ Meteorol Mund 48(1):16–21

Ministry of Foreign Affairs, MINTUR (2003) http://www.cubaminrex.cu/English/Look_Cuba/Economy/economy_Ministry%20of%20Tourism%20.htm Accessed 3 Dec 2010, Republic of Cuba

Oro JR (1992) The poisoning of paradise: the environmental crisis in Cuba. Endowment for Cuban American Studies, Miami

Press L (2011) The state of the internet in Cuba, active link: http://laredcubana.blogspot.com/2011/03/cuba-needs-domestic-upgrade-to-utilize.html. Accessed 3-2011

Whittle DJ, Lindeman KC (2004) Protecting coastal resources in Cuba; a look at current laws and institutions active link: http://nsgd.gso.uri.edu/riu/riuc04001/riuc04001_part3.pdf

Whittle D, Santos R (2006) Protecting Cuba's environment: efforts to design and implement effective environmental laws and policies in Cuba. In: Pérez L Jr, Stoner KL, Perez GMG (eds) Cuban studies, vol 37. University of Pittsburg Press, Pittsburgh, pp 90–103

Whittle D et al (2003) International tourism and protection of Cuba's coastal and marine environments. In: Parr T, McLand D (ed) Tulane Environ Law J 16(Special issue):533–579

Chapter 4
Establishment and Description of Current Park/Protected Areas System

J. Gebelein, *A Geographic Perspective of Cuban Landscapes*, Landscape Series 15,
DOI 10.1007/978-94-007-2406-8_4, © Springer Science+Business Media B.V. 2012

Abstract Chapter 4 describes the laws that define the current system of parks and protected areas in Cuba. There are different levels of protection in Cuba depending on designation whether terrestrial or marine. The maintenance of each protected area also varies in accordance with its place in the preservation hierarchy. Finally the chapter concludes with an illustration of some of the interesting flora and fauna protected in several of these conservation areas.

Keywords Parks • Protected areas • Management categories • IUCN • Conservation • EDF • UNEP • SNAP • CITMA • Article 89 • Law 81 • National System of Protected Areas • Law 201 • UNESCO MAB Biosphere Reserves • World Heritage Sites • Ramsar Sites • Ramsar Convention • Article 5 • SAMP • Marine protected areas • Law 212 • Coastal Zone Management

4.1 Introduction

Cuba has one of the highest levels of biodiversity in the world with a very large number of endemic species (Niekisch and Wezel 2003). The need for protected environmental areas in Cuba is like any other nation with its precious natural resources. The largest threat to these diverse flora and fauna is from a loss of natural habitats. In the case of Cuba, the exploitation of the lands for the purpose of mining, sugarcane, other agriculture, depletion of soil over time, logging efforts, cattle grazing and urban expansion are all cause to carefully identify these important habitats or unique and delicate ecosystems and establish legal protection.

The island has many parks, preserves, conservation areas, and world heritage sites. As described in Chaps. 2 and 3, there were many efforts from the time of the Spanish rule to preserve the forests and natural lands from over consumption. However these laws and policies were largely put into place due to competing interests rather than an interest in species conservation. For example in 1550 when the Havana city council made it illegal for slaves to cut down mahogany and cedar trees inside a radius of 8.5 km from the city. The slaves had historically used these trees for building rudimentary boats. The purpose of this restriction was to guarantee an adequate quantity of wood for the Catholic Church and for houses in the city. A second example of these early restrictions is when the Havana city council approved measures to restrict *forasteros* (outsiders) and shipbuilders from logging trees and transporting the wood to Castile to sell or utilize for their own building projects (Monzote 2008). These two examples show that it was not an interest in conservation of the land that spurred these measures per se, but an interest in saving those resources to be exploited at a later date by more 'appropriate' parties. However these two examples also illustrate that for whatever reason, there *were* legal protective measures for forests and other natural resources well before the Twentieth century. Chapter 1 details the levels of natural resource loss over time.

Since those times Cuba has established an intricate system of protection for both terrestrial and marine lands. There are many agencies that were established to determine, monitor and protect unique ecosystems throughout Cuba. Due to the efforts of those agencies in partnership with scientists, NGO's and larger worldwide environmental organizations such as the United Nations Environmental Program (UNEP) and the Environmental Defence Fund (EDF); protected areas in Cuba have been established and are consistently being updated and enlarged to try and match the changing demands on Cuba's fragile ecosystems.

Since the early 1900s the number of these protected areas has risen dramatically as awareness of the total number and diversity of flora and fauna has become more widely established. Although there has been an increase regarding the level of protection of many species of plant and animal, there have also been increasing levels of complexity pertaining to the governmental agencies that manage these protected areas.

Before the management and structure of the various agencies are discussed, it is important to note that there has been a lot of misunderstanding regarding the definition of a Cuban park or preservation area. During my travels and research of Cuba it is common to read about or personally encounter small reserves, botanical gardens or animal collections located inside resort properties that use the word "park" or "preserve" in their marketing materials. Sometimes this information is not clear until one actually arrives at the "park" only to discover it is a small private enterprise with depressingly small cages and nowhere for animals to hide and sad specimens of natural vegetation examples. Thus these smaller scale enterprises continue to exist throughout Cuba and are clearly not under any government control or subject to any regulatory policies. However, these establishments appear to be in the minority and there are many acceptably regulated and well managed areas in Cuba. There have been many complaints that Cuba's national park, preserve and reserve systems are complex and difficult to comprehend. For this reason this chapter exclusively focuses on these complex categories and designations and hopefully brings some clarity to the reader. Cuba has so many different ecosystems, elevations, soil conditions and vegetation types it is easy to comprehend why there needs to be a complex set of agencies to manage those varied types of environments.

4.2 Establishment of Protected Areas and History

The Cuban government began establishing formal designations of distinctive natural areas in the 1930s beginning with the first national park: Parque Nacional Pico Cristal. Since that time Cuba has set up a very large, and very complicated network of marine, marine-coastal, and terrestrial protected areas. Of all the different ecosystem types that exist throughout Cuba, officials estimate that almost 98% of them are represented in the existing National System of Protected Areas (SNAP). SNAP is not one single entity but truly a system that functions because of the integrated efforts of many institutes, centres, agencies and ministries that govern Cuba's marine and terrestrial resources. The development of this protected areas system was established via a series of steps described here. Law 81, also called the Law of the Environment, established the legal framework for the National System of Protected

Areas in 1997. This law states that CITMA is the supervisory lead for protected areas and leads the management, oversight and overall administration of the National System of Protected Areas under Article 89 of Law 81. Within Law 81, Article 90 lists 13 goals for the National System of Protected Areas which are described (paraphrased from original for the sake of brevity) in short below:

1. To preserve representative samples of each region of Cuba, particularly those areas that are important to migratory species.
2. To conserve the current flora, fauna and biodiversity overall, and protect it from any action or omission that could negatively impact it.
3. To ensure that reasonable techniques for local agricultural production are implemented that reach a manageable balance between raising the socio-economic level of local populations and conservation/use practices that could potentially effect delicate ecosystems such as wetlands, mangroves, karst formations, island groups, and mountainous areas.
4. To manage, protect and restore coastal and marine natural resources for conservation yet successfully manage them for sustainable use.
5. To maintain a successful balance between maintaining and managing biotic resources, providing essential goods and services to the Cuban population and making informed decisions regarding national and international regulations of terrestrial and aquatic resources.
6. To establish and maintain good practices of controlling acidification, salinization, sedimentation, restoration of soils and erosion control.
7. To carefully consider watershed management in an effort to conserve and manage water resources.
8. To strike a successful balance between forest management for the purpose of sustainable production of forest products versus improvement of forestry resources for environmental regulation.
9. To preserve the connection between the natural surroundings and cultural and historical ideals.
10. To safeguard and restore cultural and natural landscapes.
11. To encourage active involvement of local populations with nature by developing environmental education programs.
12. To achieve a successful balance between tourism and recreational development and conservation management of any area.
13. To function as a logical framework and natural laboratory for research.

(*Above are abbreviated descriptions of the 13 goals from Article 90 in: Decree-Law No. 201, translation published in 1999*)

After 4 years of negotiations, in 1999 the Cuban government ratified Law 201 to lay the groundwork for the identification, proposal, management, and creation of protected areas. And equally important, this law created the administrative hierarchy for the National System of Protected Areas (SNAP), divided the protected areas into eight unique categories, established planning requirements, and instituted precise guidelines for recommending and approving new protected areas. For example, in addition to the creation of SNAP, the Cuban government has signed international

Fig. 4.1 This map shows the current (as of 2011) UNESCO biosphere reserves, natural world heritage sites and the Ramsar wetland sites

agreements to expand conservation measures to other world-wide defined types of protected areas such as Ramsar sites, World Heritage Sites and Biosphere Reserves As of 2011, there were six UNESCO Biosphere Reserves: Buena Vista (declared 2000 with 313,500 ha), la Ciénaga de Zapata (declared 2000 with 625,354 ha), Cuchillas del Toa (declared 1987 with 208,305 ha), Baconao (declared 1987 with 92,360 ha), Guanahacabibes (declared in 1987 with 119,189 ha), and Sierra del Rosario (declared in 1984 with 25,000 ha). In this same time period, there was the incorporation of six major wetland areas onto the list of wetlands of international importance of the Ramsar Convention. These Cuban wetlands included Gran Humedal Norte de Ciego de Avila (declared 2002 with 226,875 ha), Buena Vista (declared 2002 with 313,500 ha), Humedal Delta del Cauto (declared in 2002 47,836 ha), Ciénaga de Zapata (declared 2001 with 452,000 ha), Ciénaga de Lanier y Sur de la Isla de la Juventud (declared in 2002 with 126,200 ha), and Río Máximo-Caguey (declared 2002 with 22,000 ha). The total area of these combined wetlands was 1,188,411 ha. The World Heritage Sites in Cuba are currently the Parque Nacional Desembarco del Granma (declared 1999 with 32,576 ha) and Parque Nacional Alejandro de Humbolt (declared 2001 with 69,341 ha) (Decree-Law No. 201 1999). These combined globally recognized areas are shown below (Fig. 4.1).

4.3 Different Levels of Protection (and How Are They Designated?)

The SNAP has evolved to include more protected areas over time. These management categories are defined by Article 5 in Law 201 within the National System of Protected Areas. Each one of these categories has been assigned a different conservation type by the International Union for Conservation of Nature (IUCN). The IUCN is the largest global consortium of nongovernmental and governmental organizations focused on conservation issues.

These categories are very similar for marine, coastal and terrestrial areas. The categories are listed below:

Management categories	Number
Natural reserve	8
National park	14
Ecological reserve	23
Outstanding natural element	4
Fauna refuge	10
Managed flora reserve	11
Protected natural landscape	2
Protected area of managed resources	8
Total as of 2002 (Domech and Glean 2001):	80

These eight management categories are described in detail on SNAP's website, which is the most current version of these categories. But they have also been described in detail in the following reference documents: Domech and Glean 2001; CNAP 2004; CITMA 1999; and http://www.snap.cu/. An abbreviated description of these eight management categories is given below.

Natural reserve (UICN Category I): The natural reserve is an uninhabited marine, marine-coastal or land-based area that is intended to house rare and endangered fauna or flora. It requires strict protection measures.

National park (UICN Category II): The national park is a land-based, coastal or marine area with little to no human habitation. Its design is to protect and conserve one or more unique ecosystems.

Ecological reserve (UICN Category II): The ecological reserve is a land-based, coastal or marine area that is designed to protect and conserve ecosystems or part of ecosystems. This differs from national parks because ecological reserves do not contain complete ecosystems and also can contain natural or semi-natural (human-altered) environments.

Outstanding natural element (UICN Category III): The outstanding natural element is a region that has at least one type of unique terrain feature that is not man-made. It is usually a feature that is unusual enough to have established itself in the Cuban culture in some way as to be a permanent part of cultural history.

Managed Flora Reserve (UICN Category IV): The managed flora reserve can be natural or a mix of natural and manmade land that needs regular attention to sustain its flora species and habitats these plants require to live. Its purpose it to provide an ideal environment and appropriate protections for the vegetation to grow and thrive.

Fauna Refuge (UICN Category V): The fauna refuge is a land-based, coastal or marine area that is home to some species of wild animal that requires that space either as a permanent resident or migratory species. These areas can be natural or partly non-natural.

Protected Natural landscape (UICN Category VI): The protected natural landscape is a land-based, coastal or marine area that is natural or partly natural and can be used for eco-tourism and other purposes. These are not pristine or fragile environments but tend to have some representative species of the area in that they represent.

Protected Area of Managed Resources (UICN Category VII): The protected area of natural resources is a land-based, coastal or marine area that houses a significant part of the biological diversity of one or more resources throughout the island. Concurrently, these types of areas can satisfy local or national requirements for sustainable tourism. Additionally, depending on their national or international significance, these areas can receive other formal designations as well, especially if they have some cultural, historical or natural value that is unique. These additional designations could be one of the following: Biosphere Reserve National Monument, World Heritage Sites or Ramsar sites.

The above list is a condensed version of that information found in the aforementioned cited references with special attention to CITMA 1999.

While this list and descriptions of Cuba's Management categories is helpful, there is a more general and inclusive governmental structure called the "National System of Protected Areas" (SNAP) that organizes the land and waterscapes into three overarching categories. The first is the **"Protected areas of national significance" (APSN-*Spanish acronym*)** which is defined as a region that is somehow important to Cuba on a national as well as international level and has some degree of conservation efforts that are showcased in that protected area. The second is the **"Protected areas of local significance" (APSL-*Spanish acronym*)** which usually has some historical or cultural importance and while not as significant on a national level, is quite noteworthy to nearby residents and usually part of the local history in some way. The third level is the **"Special regions of sustainable development" (REDS-*Spanish acronym*)** which tend to be large sections of Cuba's territory which contain delicate ecosystems that are of cultural and economic importance and these areas are seen as examples of areas with the topics of sustainable development and conservation as their focus of design.

These two sets of management hierarchies are helpful in giving context and background to Cuba's complicated protected areas legislation and approach to preservation and conservation of the varied landscapes in its territory both terrestrial and marine. The basic laws and descriptions of designations for various categories of protected areas are presented here. Most of these grew from an appreciation of the land-based and coastal ecosystems and habitats. Since this was the case it is helpful to now discuss the evolution of Marine Protected Areas (MPAs) in Cuba.

4.3.1 Marine and Coastal Protected Areas

The marine and coastal protected areas in Cuba are managed by the Cuban Subsystem of Marine Protected Areas (SAMP). This subsystem is under the administrative umbrella of the National System of Protected Areas. The marine protected areas (MPAs) are defined as those areas that contain a marine or coastal component within their boundaries. This includes offshore keys, the submerged coastal zone and coastal wetlands. Their evolution to present day consisted of several workshops where areas were proposed to be assigned a "protected area" status. SNAP first received these proposals from 1968 to 1973. However these first proposals were largely terrestrial based with small numbers of lagoon, mangrove and coastal ecosystems included. FAO consultant Kenton Miller was the lead on their development (Miller 1984; CNAP 2004).

The establishment of Marine Protected Areas (MPAs) was slower than the terrestrial protection system. For example the first National Protected Areas Workshop in 1989 received proposals for protected areas throughout Cuba but while there were preliminary discussions concerning the potential for marine parks, the land-based and coastal foci dominated the workshop. However in 1995, the second National Protected Areas Workshop made marine parks and preserves its own subdivision of SNAP. Numerous marine parks and preserves that are part of the current SAMP date back to this second workshop. The importance of marine preserves evolved into a program of work in the third National Protected Areas Workshop in 1998. Both meetings and courses were held from 2000 to 2001 with help from the Global Environment Facility (GEF) and the United Nations Development Program (UNDP). These meetings and courses ultimately led to the formation of the 2003–2008 SNAP Plan. In the SNAP Plan major concerns were described, ranked and a system of regulations for conservation and use of marine ecosystems was established. Some of the major concerns included pollution prevention and reduction, demarcation of SAMP areas, and re-establishment of habitats (CNAP 2002). The overall success of the SAMPs can be measured in the number of MPAs that have been approved and those that are in the process of being approved. In 2008 it was estimated that 108 MPAs were approved. This number of MPAs comprises nearly 11% of the ultimate goal of preserving 25% of the Cuban marine bed to a depth of 200 m. Of these 108, almost half (49) are established nationally significant preserves and the rest have only local importance.

The legal definition of the SAMP is based on the terrestrial system and there are no true legal distinctions between the two systems. In fact the SAMP is part of SNAP since in 1997 Law 81 defined SNAP as an integrated marine-terrestrial system. A multinational effort ensued during this 2003–2008 time period and included participation by the Environmental Defense Fund (EDF), WWF Canada and a large number of the Cuban scientific organizations and managers of protected areas. The result of this effort confirmed the importance of GIS, digital cartography, remote sensing, decision support systems, ecoregional planning, habitat classification, and spawning aggregation sitings in the strategic design and administration of current and proposed MPAs in Cuba (Sala et al. 2002; Airame et al. 2003; Leslie et al. 2003; Roberts et al. 2003).

The coastal zone, defined by Law 212, was the last major category of Cuba to be protected. There was an interesting transition period for this law in that before 1997, the first iterations of this bill were focused on trying to control both occupation and construction in the coastal zone. With the approval of Law 81 (Law of the

Environment) and after 1997, the legal structure protecting marine and coastal areas began to embrace more specific regulations that restricted marine fisheries, began the process of establishing an environmental impact assessment on these areas, and drafted a series of repercussions for environmental violations (CITMA 2000).

The Council of State of the Republic of Cuba accepted Law 212 called "Coastal Zone Management" on August 8, 2000. This law detailed the scope and size of the coastal zone to guarantee its protection and sustainable use. The reasons this law was created included: Erosion and anthropogenic impacts linked to fishing, tourism, various industries, agriculture and transportation (CITMA 2000). This law was anticipated to protect coastal regions from negative environmental impacts correlated with these and other anthropogenic activities. Law 212 explicitly defines two zones: the "zone of protection" and the "coastal zone". The "coastal zone" stretches from the high water mark inland for a distance of 20–40 m. This distance discrepancy depends on the type of coastline (steep, flat, rocky, mangrove, et cetera). Permanent structures are usually prohibited in this zone unless justified for social well-being or public utility. Residences, waste disposal sites, sand extraction, most motor vehicles, new hotels and associated activities are expressly prohibited.Bordering the "coastal zone" is another protected area named the "zone of protection". This boundary expands inland an additional 20–40 m from the external boundary of the "coastal zone". The main differences between the two are that the "zone of protection" allows for select light construction of non-permanent structures for example some types of concession stands. Secondly, the "zone of protection" allows for agricultural crops to be grown as long as these crops do not threaten to replace natural vegetation or impact the ecosystem in any negative fashion. More specialized regulations for small keys and islands are provided by Law 212 including the prohibition of any new construction on any key or island that is entirely covered by the zone, development of beaches, removal or alteration of mangrove vegetation, or an area that is designated fragile because of its geomorphological development. Law 212 is an excellent step towards protecting the coastal environment from tourism because of the setbacks just described, this buffer area will, if enforced, be an extremely effective conservation technique to preserve these sensitive waters, dunes, wetlands and mangrove ecosystems (CITMA 2000).

4.3.2 *Differences Between Land and Marine-Based Management Protection System*

In the early stages of protected area selection, design and some elements of management the same guiding principles can be following whether the target area is terrestrial or marine-based (Estrada and Perera 1998; Kelleher 1999; PNUMA-PAC 2000). Going further with this point, even though the legal framework for marine and terrestrial environmental preservation is based on the same Decree Law 201, there are some differences between the two in terms of management and planning knowledge and standards. Some of these contrasts include: these areas are easily accessible; management necessitates a larger financial investment; marine biological knowledge is

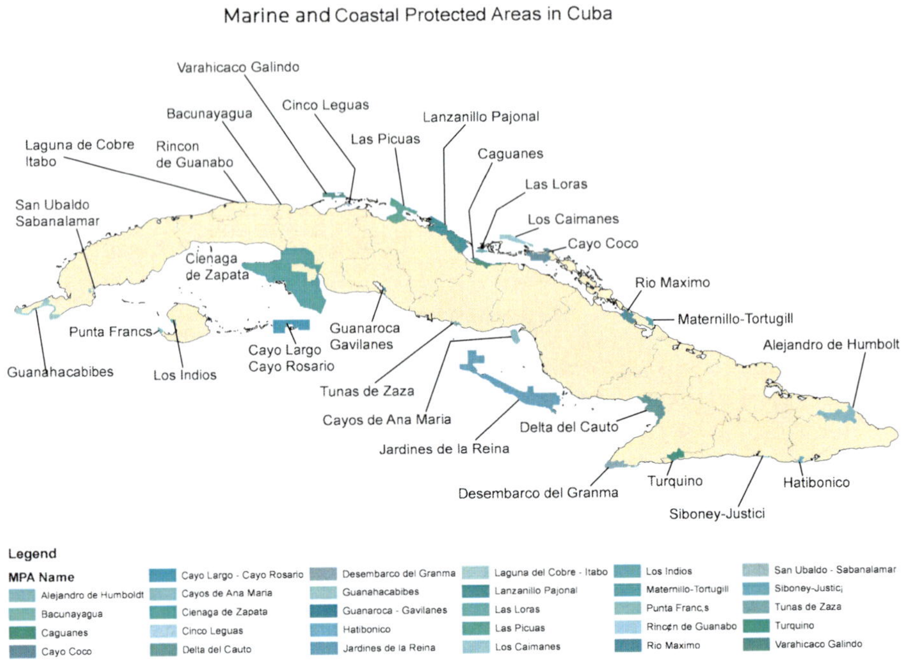

Fig. 4.2 GIS map of MPAs surrounding Cuba

not as well known as land-based ecosystems, the MPA's have a relatively high use compared to terrestrial protected areas, and the ocean is considered a public resource whereas many land-based protected areas in Cuba are not. One such public resource is the extensive coral reef formations that have been utilized to create more meticulous conservation zones. It is interesting to note that because the basis for MPAs was derived from terrestrial protected areas, many times multiple ecosystems were included in the same MPA type to fit a landscape ecology model. Based on these design methods some MPAs extend out from existing land-based protected areas to the ocean and shallow waters near the coast (CNAP 2004). It is obvious that this is the case when looking at a GIS map of the MPAs shown below (Figs. 4.2 and 4.3).

Another significant difference between the two exists regarding zonification. One of the key components of establishing the various MPA zones are the lines of living coral that inhabit the waters off of Cuba's coastline. Additionally, the varying bottom depths and types including blue holes and banks incorporate different ecosystems that do not neatly fit into the legal protection of the terrestrial structure. There are also socioeconomic use zones that translate into different resource uses like the different methods of fishing such as commercial fishing, catch and release, or drag netting. Those three techniques of fishing can have completely different impacts on fish populations and thus are treated differently by zonation. Some zones, because of the different pressures of fisheries techniques, overlap depending on time of year and species of fish. In some zones there are "no take" rules where the only possible acquisition of marine species is allocated to scientists (CNAP 2004).

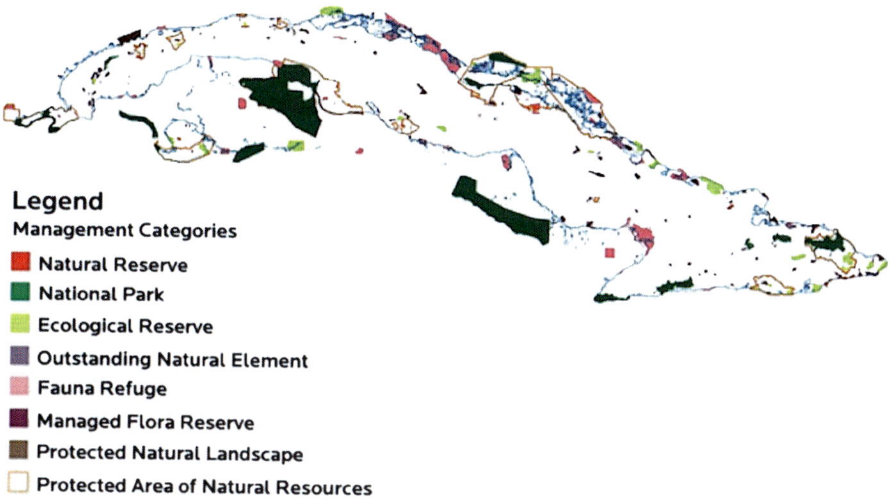

Fig. 4.3 Managed areas map (this is an adaptation from the map shown here: http://www.snap.cu/html/snap.htm)

4.3.3 How Are Parks (or Any Protected Areas) Maintained?

The maintenance of a protected area in Cuba varies widely. The level of maintenance is highly dependent on the resources given to the managing agency or institute such as resources to hire naturalists or GIS experts, boats for patrolling marine areas and conducting research, gas for any motor vehicles (terrestrial or marine), acquisition and maintenance of field equipment, wet lab infrastructure and equipment, flora and fauna inventory equipment, for example. If the total number of park employees is 20 for a reserve the size of the Zapata, the maintenance of that national park will be quite limited because of the large size of the park.

A second factor impacting park maintenance is understanding current threats to habitat biodiversity or individual species. For example in Cuban commercial fisheries there are three main species that are taken: the queen conch (*Strombus gigas*), two shrimp species (*Litopenaeus schmitti* and *Farfantepaeus notialis*) and the spiny lobster (*Panulirus argus*). There is certainly anthropogenic pressure on these species, but the amount taken is strictly enforced by the government and even with the small level of sport fishing allowed, the waters are not yet depleted. However another example in the 1950s in the Gulf of Batabano the Biajaiba (*Lutjanus synagris*) was completely depleted by overfishing because of a lack of sustainable governance at that time (Ortiz and Lalana 2008). The establishment of limits on terrestrial and marine habitats and ecosystems requires scientific research to be planned, acquired, analyzed, and reported to the governing body of any protected area. Therefore it is common for many environmental agencies to partner with universities and centres throughout Cuba to tap into the regional scientific expertise such as National Centre of Biodiversity (CeNBio), the National Centre for the Protection of the Flora and

Fauna (E.N.P.F.F), the National Botanical Garden (JBN) or the Institute of Ecology and Systematics (IES). It is also common to reach out to international agencies such as Ramsar affiliated scientists, the United Nations Environmental Program, or researchers of the Environmental Defense Fund (EDF) for help and collaboration regarding protected area rationale and delimitation.

A third factor is the infrastructure that exists for the management of protected areas. By this I mean the day to day, month to month and year to year operations. This varies widely, even within the same protected category. The Ciénaga de Zapata and the Guanahacabibes are both UNESCO Biosphere Reserves. However, the management of each is quite different. The Guanahacabibes peninsula is quite remote (particularly on the mid to far western end), consists of a rocky shoreline populated with mangrove, some sandy beach, and oddly, stunted plumaria trees on the south eastern side. There is one road in and out and, depending on the time of year, it can be completely or partly washed out and impossible to use for access to research sites. There is no gate, the road there is largely unmarked with any signs, and there is no associated nature centre on site with the exception of a tiny building for the naturalists who live in that area. When I personally visited this site I had to make inquiries in Santa Maria, the small town nearby, to find someone who would accompany me out into the reserve. Thus, in this case, maintenance is minimal; there is very little day to day staff, and only a handful of naturalists who are available on an irregular basis. In one sense this situation is excellent for the preservation of migratory birds who use this area as a stopping point when travelling north or south, there is little human disturbance. And secondly, because of the minimal maintenance of this reserve, there is very little impact on this landscape from tourism since it is remote and fairly inaccessible for the casual tourist. The Ciénaga de Zapata is entirely different. It has several populated areas within a day's drive and has a much larger full time staff of naturalists, scientists and park care-takers. The Ciénaga de Zapata also has a much higher tourism infrastructure with several buildings, well paved roads, observation points throughout the preserve, and many hiking trails that can be used by tourists or scientists (but either must be accompanied by a Zapata guide). It has a worldwide reputation as being home to the Cuban crocodile, and hundreds of bird species who migrate through the swamp every year, including the pink flamingo. These two examples exemplify the vast differences in maintenance and conservation approach to reserves in the same category. There are many other parks and protected areas that also contrast greatly in their care, tourist impacts, overall public visibility and access for scientific purposes.

4.4 Examples of Native/Exotic Species of Flora and Fauna

Cuba has every type of ecosystem and landscape in the Caribbean represented on its surface. To showcase that variability the maps below and associated images depict some of Cuba's unique and most interesting landscapes, flora and fauna.

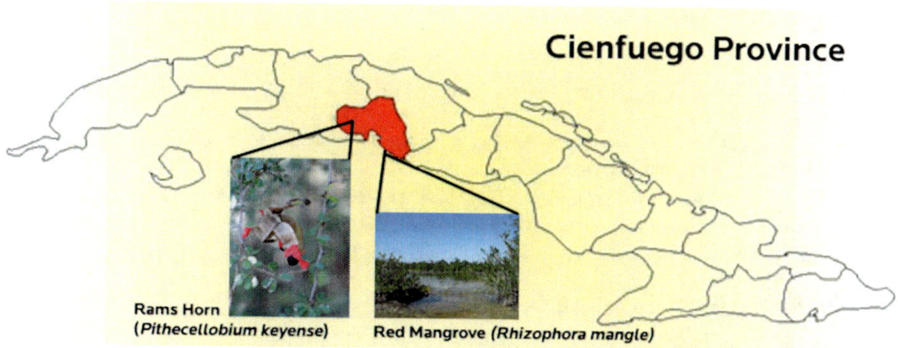
Cienfuego Province
Rams Horn
(Pithecellobium keyense)
Red Mangrove (Rhizophora mangle)

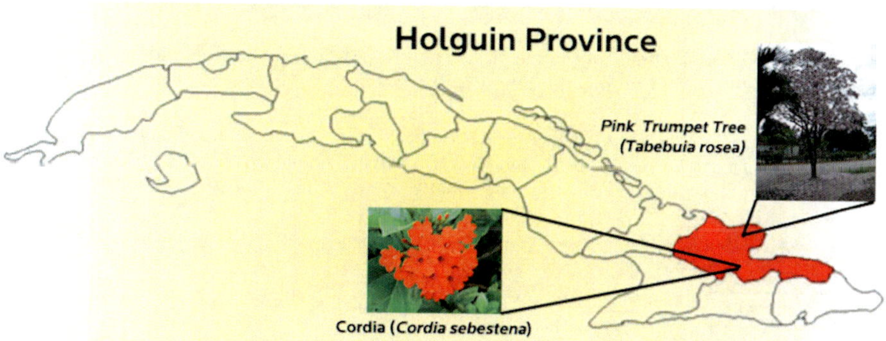
Holguin Province
Pink Trumpet Tree
(Tabebuia rosea)
Cordia (Cordia sebestena)

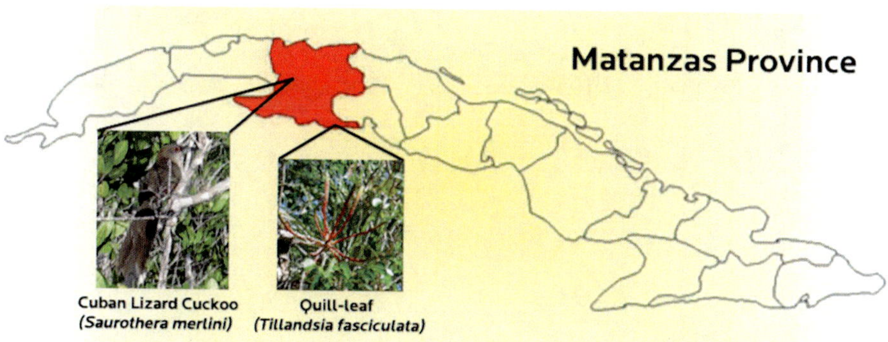
Matanzas Province
Cuban Lizard Cuckoo
(Saurothera merlini)
Quill-leaf
(Tillandsia fasciculata)

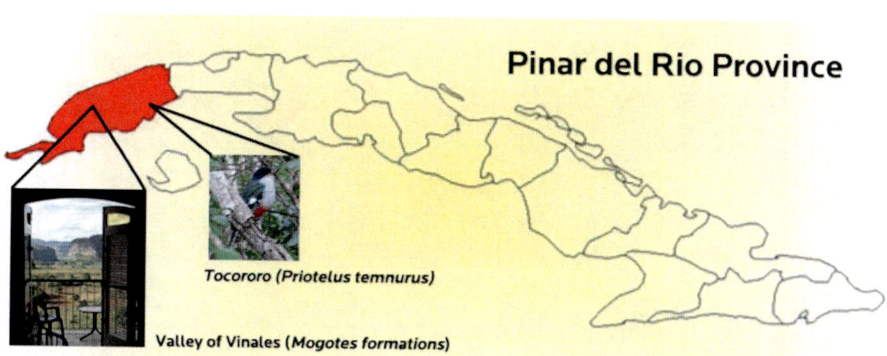
Pinar del Rio Province
Tocororo (Priotelus temnurus)
Valley of Vinales (Mogotes formations)

References

Airame S et al (2003) Applying ecological criteria to marine reserve design: a case study from California Channel Islands. Ecol Appl 13(1):475–510

Centro Nacional de Areas Protegidas (CNAP) (2002) Sistema Nacional de Áreas Protegidas, Cuba. Plan 2003–2008. CNAP. 202 pp

Centro Nacional de Areas Protegidas (CNAP) (2004) The national system of marine protected areas in Cuba. National Center for Protected Areas, Habana

CITMA (1999) National system of protected areas, Decree-Law No. 201. Environmental Law in Cuba Series

CITMA (2000) Coastal zone management, Decree-Law Number 212–2000. Environmental Law in Cuba Series, Habana

Domech R, Glean M (2001) Cuban nature. Publicaciones en languas extranjeras, La Habana

Estrada RE, Perera A (1998) Determining significance of protected areas in Cuba. Int J Wilderness 4(2):13–16

Kelleher G (1999) Guidelines for marine protected areas. IUCN, Gland/Cambridge, 107 pp

Leslie H et al (2003) Using siting algorithms in the design of marine reserve networks. Ecol Appl 13(1):185–198

Miller K (1984) Planificación de parques nacionales para el ecodesarrollo en latinoamérica. Fundación para la Ecología y la Protección del Medio Ambiente (FEPMA), Madrid, 500 pp

Monzote R (2008) From rainforest to cane field in Cuba. University of North Carolina Press, Chapel Hill

Niekisch M, Wezel A (2003) Protected areas in Cuba – developments and problems. Natur und Landschaft 78(8):360–366

Ortiz M, Lalana R (2008) Marine biodiversity of the Cuban archipelago: an overview Center for Marine Resarch. University of Havana, Havana

PNUMA -PAC (2000) Manual de capacitación en el manejo de areas marinas protegidas. Bayahibe, República Dominicana. 63 pp

Roberts CM et al (2003) Application of ecological criteria in selecting marine reserves and developing reserve networks. Ecol Appl 13(1):215–228

Sala E et al (2002) A general model for designing networks of marine reserves. Science 298:1991–1993

Chapter 5
History of Remote Sensing and GIS as It Relates to Assessment of Land Use and Land Cover Changes Over Time

Abstract This chapter describes the technical history and detailed background of remote sensing and Geographic Information Systems (GIS) both in the broader discipline as well as how they are relevant and important to Cuban scientific research. First an abbreivated history of remote sensing and GIS is given and then how this relates to earth observations in the Caribbean and Cuba. Several analyses are shown as examples illustrating impacts on the Cuban terrestrial and marine territories such as threats to marine areas from urban growth and indirect impacts from pollution issues such as sewage and sediment deposition.

J. Gebelein, *A Geographic Perspective of Cuban Landscapes*, Landscape Series 15,
DOI 10.1007/978-94-007-2406-8_5,

Keywords Mercedes (land grants) • GIS • Remote sensing • Photo-interpretation • Digital image processing • Space station • Tamayo Méndez • MIR • ISS • NASA • DoD • ESRO • ESA

5.1 Introduction and Historical Context

There have always been ways of assessing a landscape's anthropogenic development and natural evolution. Understanding *how* Cuban landscapes were assessed gives us insight into the perception and value of such landscapes in the eyes of early explorers up through modern day developers. Documented landscape change is a written record of an

> ...expression of the dynamic interaction between natural and cultural forces in the environment. Cultural landscapes are the result of consecutive reorganization of the land in order to adapt its use and spatial structure better to the changing societal demands. (Antrop 2005)

So clearly landscape change is in some sense an expression of the dominant culture at any given time. Landscape change usually begins with an evaluation of the land and its monetary or cultural value. Such assessments of land have utilized certain methodologies or technologies which have changed dramatically from the times of Columbus to present day. During the early European history in Cuba there were surveying techniques which initially relied heavily on the knowledge of local indigenous peoples and their trails throughout interior the island. The initial surveys of Cuba resulted in the land being divided grant (*mercedes*) in the shape of a circle. This method was based on a sixteenth century method by the Spanish Crown to give land to their followers in the Caribbean. Unfortunately, many of these grants were very vague in their delineation. Making matters worse, they were drawn without regard to forests, meadows, steep hillsides, and many times appeared to overlap into areas which were considered open for common use. Thus the different categorizations of land (such as forest, mangrove, cropland, meadow) were not considered because these categories were inside the allocated lands, and what was deemed important was the ability of the landowner to provide beef to inhabitants of the nearby settlement and house an inn at his land's centre (Thomas 1998). Since these territories were difficult to define from the time of Columbus, this makes it quite challenging to establish exact changes in cultivated areas. However, as noted in earlier chapters, there exist sufficient records of exports and consumption of various resources on the island, from which it can then be determined the amount of land that was altered due to logging, tobacco and cane cultivation, for example.

Geographically the shape of these *mercedes* is still detectable on Cuba's landscape today. Upon examination of any detailed twentieth century map shows how many city boundaries, farm roads have been established based on these early circles. Many town centres are loosely laid out as a circle (Thomas 1998).

Generally speaking a cartographer, sketch artist, and naturalist were taken on all exploratory voyages where land was to be claimed and maps needed to be created for the purpose of future harbour establishment, locations of freshwater springs, major and minor rivers, land for houses, prominent natural resources such as forests, and promising locations to search for gold and other precious treasure. The exception to this, of course are the island and mainland coastlines, they were the most easily accessible and therefore in early maps are the most detailed in regard to geography, topography, bays, harbours and inlets.

Currently there are multiple ways of evaluating both anthropogenic and natural landscape changes. The first and one of the most effective methods for assessing multiple spatial scales is remote sensing. Since the major thrust of this chapter focuses on technological developmental assessment, we will first discuss remote sensing and secondly Geographic Information Systems (GIS) and the application of both in landscape change over time in the subsequent chapter sections. First remote sensing is defined, secondly a brief history of remote sensing is described in the context of the United States, Europe and the Soviet Union combined with Cuba. Then we discuss what makes a good analyst, touch on the technology that acquires this data and finally come full circle with how this data is transformed into meaningful information. Geographic Information Systems (GIS) is then also defined, historically described in the same context, and finally the acquisition technology, expertise required and translation of data into information is presented.

5.2 Remote Sensing

Remote sensing is the practice of acquiring information about something without being in direct contact with it. Lillesand et al. (2008) have a much more precise definition:

> *Remote Sensing* is the science and art of obtaining information about an object, area, or phenomenon through the analysis of data acquired but a device that is not in contact with the object, area, or phenomenon under investigation.

This is in contrast to *in situ* sensing which is the opposite type of data acquisition; gathering data about an object or phenomena while being in direct contact with it. For example a measurement of water temperature or pH of soil requires that an instrument be in direct contact with water or soil. *In situ* sensing is often combined with remote sensing data to verify or validate assumptions made in remote sensing calculations. This will be discussed in more detail in the applications section of this chapter. To put in context the evolution of remote sensing and how crucial it can be to assess both single incidents in time as well as document and identify changes, a brief history of remote sensing is presented.

Remote sensing had its beginning in photography. The ground-breaking photographic techniques of Nicephore Niepce, Louis Jacques Mande Daguerre and William Henry Fox Talbot were revealed to the public in 1839. One year later the director of the Paris Observatory encouraged the use of photography for topographic surveying. The history of studying the earth from above begins with aerial photography. Gaspard-félix Tournachon took the first aerial photograph of Paris from a hot air balloon in 1858 from a height of 1,200 ft above the city. American James Wallace Black followed soon after with the first aerial photograph on the North American continent in 1860 over Boston. During the 1860s aerial photography from balloons was common to obtain information about forested terrain and enemy movements during the American Civil War. Following the American War, in 1887 the Germans initiated research using photography to delineate forested areas from other types of land cover and land use. This research led to photogrammetric methods for measuring areas and features (Gebelein 2003; Lillesand et al. 2008). Balloon photography evolved into kite photography in order to study meteorological phenomenon beginning in 1882. In fact the first kite photograph is credited to a meteorologist named E.D Archibald. In 1890 Frenchman Arthur Batut published La "photographie aérienne par cerf-volant" (Aerial Photography by Kite), that discusses different targets of interest including agriculture, urban areas, archaeology, and aerial reconnaissance. Before airplanes there was the "photo rocket" created by Amadee Denisse in 1888 which launched a camera into the sky and the user then retrieved the camera system. Alfred Maul patented a system in 1903 whereby a camera attached to a rocket achieved heights of 2,600 ft. Further experimenting with different platforms took an unusual twist in 1903 with Julius Neubronner's breast-mounted aerial camera made for carrier pigeons. It weighed 2.5 ounces and the film was exposed every 30-seconds from this system. In 1906 G.R. Lawrence showcased photographs of the aftermath of the terrible San Francisco earthquake and that documentation received worldwide attention. This was one of the first instances of aerial photography being used to show the effects of damage after a natural disaster. In 1909, the first aerial photograph was obtained from an airplane over Centocelli, Italy by Wilbur Wright. Aerial photography during World War I was done initially with the camera attached to the chest of the cameraman, or (later) to the side of the plane. Aerial photography became an integral part of World War I reconnaissance for identifying usable versus damaged barriers, roads and vulnerable points for the movement of troops (Gebelein 2003; Jensen 2000; Lillesand et al. 2008; Fischer 1975). Aerial photographs also provided a new prospective of the enemy's equipment and troops, many types of military materials were easy to conceal from the ground perspective but impossible to conceal from the air. The height of aerial photographic processing during this period was in 1918 when French aerial units were developing and printing up to 10,000 prints per night in times of elevated activity. During World War II more advanced methods in Aerial Photographic Interpretation (API) were created. These methods incorporated detection methods for recognizing invasion barges, troop movements and damage assessment. After World War II an enormous effort began by both the United States and the Soviet Union to collect information about each other. Both sides publicly stated that "…any aerial reconnaissance overflight of another state without authorization was considered to be an illegal and hostile act unless national leaders agreed to it beforehand" (Jensen 2000). However such overflights were essential to gain insight into the other's activities. Therefore President Eisenhower launched the Genetrix Reconnaissance Balloon Project that ultimately acquired 13,813 aerial photographs of the Soviet and Chinese territory beginning in 1955. These photographs were obtained from cameras housed in these 448 weather balloons (Jensen 2000). During the next 5 years there were considerable technological advances in observational technology. Sputnik, developed by the Soviet Union, was launched in 1957 and was the world's first satellite; this officially marked the beginning of the "Space Age" (Day 1998). Another major advance was the U-2 reconnaissance aircraft that housed two high spatial resolution camera systems. The first provided a long focal length which enabled it to detect objects on the ground that were two-three feet in length from an altitude of 70,000 ft. The second camera structure acquired information on a continuous film strip for the entirety of the flight (Rich and Janos 1994). The SR-71 replaced the U2 plane with better speeds (broke world record for aircraft speed at 2,193 m.p.h.) and it also utilized new signal equipment to obtain better records of the targets below (Jensen 2000). In 1958 President Eisenhower funded the Corona Project to acquire pictures from low earth orbit of the Soviet Bloc countries. There was growing interest by the public and government entities for Earth observations. In response to this aerospace research in the United States split into two divisions. The first was the public face of aerospace research: the National Aeronautics and Space Administration (NASA) and its goal was the peaceful study of space and to accomplish essential aeronautics research. The second part of the split was the Department of Defense (DoD). This second group was responsible for research and development of military aerospace activities (Burrows 1998). The hope was that once these divisions were created, maps of the earth could be finished and benefit the public. However, most of the funding for the earth observation programs and missions originated from intelligence budgets for surveillance endeavours. Therefore the cameras, film and all photographs were immediately classified until 1995 when President Clinton signed an Executive Order that declassified the space-based intelligence systems known as the CORONA, ARGON and LANYARD missions.

The exception to this was the photographs of the earth made by astronauts. Since the early days of Apollo these photographs have usually been made available to the public and to researchers (without classified clearance). In United States history there have been eight major phases of human space flight, these include the following NASA and NASA-cooperative programs: Mercury Program (1961–1963), the Gemini Program (1965–1966), the Apollo Program (1968–1973), the Skylab Program (1973–1974), the Apollo-Soyez Test Project (1975), the Space Transportation System (Space Shuttle, 1981-present), the Shuttle-Mir Program (1966–1998), and the International Space Station Program (2000-present) (Gebelein 2003). As described by Glazovskiy and Dessinov (2000) the Soviet Union also had an astronaut observation program called "Visual Observations" and from 1974 to 1976 earth observations were carried out with hand-held camera and fixed camera system.

Western European nations also split into two major space entities: the first was the European Space Research Organization (ESRO) (that ultimately became the European Space Agency), and the second was the European Launch Development Organization (ELDO). ESRO was formally recognized in 1964 after an agreement had been signed 2 years prior in 1962. ESRO performed well with seven successful missions, all achieving orbit from 1969 to 1972, and all launched by U.S. launch vehicles, thereby starting a scientific and cooperative trend. The same year that ESA was established the ESRO was joined with ELDO.

In 1975 the European Space Agency (ESA) was formed as an organization dedicated to the exploration of space and has 18 member states, It includes a program committed to human space flight mostly through the cooperative research and development programs linked to the International Space Station. ESA science missions are headquartered in Noordwijk, Netherlands, ESA Mission Control is in Darmstadt, Germany, the Astronaut training facilities are located in Cologne, Germany, the European Space Astronomy Centre is based in Villanueva de la Canada, Spain and Earth Observation Missions are in Frascati, Italy. The establishment of ESA was at an opportune time due to the fact that budgets for both the United States and Soviet Union were significantly reduced. Thus, during the 1970s the ESA became a front-runner in space exploration as well as earth observation research and development.

Certainly there have been important collaborations between the ESA and NASA, for example one of the first was the International Ultraviolet Explorer (IUE) which was an astronomical observation platform designed to acquire ultraviolet spectra. This was considered a very successful endeavour due to the fact that the instrument was given a lifespan of just 3 years but lasted 18 years. This project included participants of the UK Science Research Council, the ESA and NASA. There have been other subsequent joint projects between NASA and the ESA analyzing outer space from earth's orbit and exploratory probes. However here we predominantly focus on earth observing missions. Beginning in the 1990s the ESA relied less on NASA for joint ventures and began collaborative projects with the Soviet Union. There were many reasons for this new partnership one of which was certainly the uncompromising legal regulations placed on the sharing of information by the U.S. military.

There were earth observing missions that were established by various countries for the express purpose of monitoring various phenomenon both episodic and over long periods of time. The best examples of such programs are the space station programs which began with the Soviet Salyut 1 which was the first space station of any kind, and was launched by the USSR on April 19, 1971. One of its top priorities was to observe the earth from space and determine challenges and identify landscape geography that was of particular interest to scientists. Following the 6 month lifespan of this first earth observation platform were many other attempts by the Soviets including Salyut 2 (launched 1973 – engine exploded after 2 weeks), Cosmos 557 (engine failed and went out of control in 1973), Salyut 3 and 4 (launched 1974), Salyut 5 (launched 1976), Salyut 6 (launched 1977), Salyut 7 (launched 1982), and the Mir station (launched in 1986).

The United States, before leading the charge with the ISS program, launched Skylab into earth's orbit on May 14, 1974 and ended its time in space on July 11, 1979. The objectives of Skylab were not earth observation-related, its purpose was purely to prove humans could exist in space for long periods of time and secondly, to increase our understanding of solar astronomy.

The Soviet Visual Observation Program was of interest to the United State's Space Exploration Program which forged the joint Shuttle-Mir earth observations partnership. This cooperative program had two major objectives. The first was to utilize earth observations of astronauts and cosmonauts to record long-term and episodic events such as fires, hurricane activity, urban growth, land cover change and floods. The second major objective was to create appropriate techniques to prepare for the next phase in earth observations from the International Space Station (Evans et al. 2000). Following the fragmentation of the Soviet Union in 1991, it was determined that no more station replacements would occur for the Mir Space Station. As an alternative the Soviet Union joined the partnership in the International Space Station (ISS) program. Mir instead became a useful testbed for ISS station procedures, equipment and experiments.

The International Space Station (ISS) was and continues to be a collaborative effort among several principle countries including: United States, Russia, Europe, Japan and Canada. The first crew boarded the (ISS) in 2000. There are numerous scientific experiments and long term monitoring programs that have occurred and that are ongoing on the ISS, including the promotion of earth observation missions focused on global environmental concerns, such as natural and human-made events on earth. Regarding global warming, astronauts have documented enormous tabular icebergs in the South Atlantic Ocean and the fragmentation of Antarctic ice shelves (Scambos et al. 2005). Crew Earth Observations (CEO) have recorded various types of human impacts since the 1960s such as agricultural growth, plankton blooms and other events requested by scientists on earth. Crew earth observations have been consistently performed since 1961 and plan to continue the program's legacy. To date over 350,000 photographs have been acquired throughout ISS operations.

Thus far the leaders in early and current space flight have been described pertaining to historical exploration into orbit and earth observations. This is useful information to give context for the technical side of remote sensing and where we are today and what we can accomplish based on these types of developments. It is beyond the scope of this book to delve into an exhaustive list of all remote sensing instruments developed, their spatial and spectral descriptions, altitudes and lifespans. However a comprehensive list and description of all these instruments and their launch platforms can be found in: Jensen (2000), and Lillesand et al. (2008).

5.3 Cuba's Participation in Space Science

Cuba has been a smaller player in the exploration of the earth from space, but it has participated. Cuba has been one of Russia's major Latin American allies for many years. One of the most famous people in this respect is Tamayo Méndez who was a member of the Cuban air force as a pilot. He was trained in the Soviet Union to fly the MiG-15 in 1961. In 1962, through the Cuban Missile Crisis he flew 20 intelligence missions. In 1978 he was chosen as a potential cosmonaut as part of the Intercosmos Program in the Soviet Union. This program approved the participation of cosmonauts from other countries (Arnaldo Tamayo Méndez 2011). In 1980 Tamayo was launched onboard the Soviet Soyuz 38 for 8 days to the Salyut 6 space station. One of his scientific contributions was to grow organic monocrystals in space using Cuban sugar. In terms of remote sensing he took multispectral photographs of Cuba's coastal waters and the submarine shelf. An interesting note about his mission was that the Cubans coordinated aerial flights to take photographs using an MKF-6 M multispectral camera system over the same areas Tamaya flew over in the station at approximately the same dates. This was the first instance of comparison between multiple elevations (orbit versus high aerial photography) with comparable instruments on similar dates in Cuba.

Cuba's space research efforts have largely been cooperative with the Soviet Union usually using Russian satellites and partnering in research projects in this manner. One of the earliest efforts was in 1964 when scientists at the Cuban Academy of Sciences began working with some of the first optical data acquisitions of the earth's surface using artificial satellites. During this same time period research was done focusing on the Doppler Effect analyzing radio signals from satellites attempting to use that extracted data for ionospheric research. In 1969 Cuba began receiving meteorological satellite imagery on the atmosphere above Cuba for aiding weather forecasting accuracy. These projects were in collaboration with the Soviet Union and Germany.

Cuba began another collaborative effort in 1975 by joining the Intercosmos Working Group for Remote Sensing to apply remote sensing techniques to study the island's many natural resources and help delineate areas in need of protection and conservation from urban expansion and development. One of the results of these initial analyses was the discovery of important geological structures and features as well as derived soil erosion maps, cropland impacted by salinity intrusion, identification of pollutants in the port of Havana, and determining areas of productive fishing zones. These were all areas of study that were aided by the remote sensing perspective and analysis. More specific applications of this remote sensing imagery are establishment of more detailed sugarcane extent maps and evaluation of unique and difficult to access areas such as the Sierra Maestra National Park.

Space-based and aerial remote sensing research and data acquisition have been slowly but steadily increasing since this first cosmonaut's launch into space. Recently there have been two major space-focused research foci in Cuba which are meteorological studies and secondly research utilizing multispectral for the purpose of updating thematic maps and charts. Geologists in particular have been using this type of imagery for mineral identification, for example. Whereas geographers have been updating maps using this imagery to more accurately portray physical landscape features such as coastal morphology; botanists are utilizing this imagery to create regional maps of vegetation types throughout the island. Finally Cuba has been strengthening its scientific activity on the subject of space science by hosting conferences and symposiums such as the First National Symposium on Space Research held in Havana in 1983 where topics included space physics, environmental earth observations, space meteorology and space laws and policies (Altshuler 1982).

Cuba has clearly exhibited an interest in space research but it is obviously not a top priority. Space research requires enormous financial resources which Cuba cannot afford to divert from its more important priorities such as energy supply, public health, agriculture, and industrial production. Now that a brief world historical context has been given for the development of space scientific efforts, we move to an abbreviated technical description of the two major research options performed on satellite imagery or aerial photography once it is acquired.

5.4 Brief Remote Sensing Technical Background

Whether for space science, intelligence work or for environmental applications remote sensing requires some knowledge of the data target as well as an understanding of the data acquisition process. Remote sensing can be thought of as a multistep process. The image must first be acquired by an instrument; this refers to the technology employed. Then the image must be processed in a way that converts the raw data into an actual image form that we would recognize. Thirdly, the image must be interpreted. This is the most important step in the process because here the image is converted into information that is meaningful and valuable. This last step is necessary whether we are conducting digital image processing using software or photo interpretation utilizing traditional mapping tools and visual analysis.

Remote sensing has changed as technology has evolved and become more sophisticated. There are two main ways of doing remote sensing: digital image processing and photo-interpretation. Photo-interpretation is the art of visually assessing a photograph in order to identify objects and evaluate their significance (Philipson 1997). Traditionally photo-interpretation was done with hard copy, print photographs developed from film. A photo-analyst would study the photograph over a light table to help bring out features of the photograph. To create 3-dimensions a stereo-pair of photographs were used to bring out depth in any landscape features along with the use of a stereo-scope. Often, clear acetate (or transparent) sheets were used to literally categorize different parts of a landscape. On each sheet of acetate paper the analyst would trace a feature type such as delineating the boundaries of forests, agricultural fields, water bodies, wetlands, or urban areas. Therefore, when giving a presentation describing the significant features of a landscape, the presenter would show, on a white background, different sheets of acetate paper illustrating different landscape categories by overlaying the sheets one by one on top of each other. The final display would reveal a colour-coded, complex landscape. Interestingly, much of what was traditionally done by visual assessment and by-hand calculations can now be done with software such as Erdas, Imagine. However even though, usually, the final step in aerial photo analysis is transferring the photograph to a digital database these aerial photographs remain a very useful part of identifying objects in a scene and judging their significance. A comprehensive description of all photographic interpretation equipment, measurement techniques and a detailed description of stereoscopic viewing can be found in Lillesand et al. (2008).

5.5 What Makes a Good Interpreter?

Remote sensing is also considered an art because it involves photo or image interpretation which largely depends on the scientific background of the interpreter as well as their life experiences. Life experience obviously varies among individuals and combined with scientific training allows an interpreter to establish rules of thumb to extract information from imagery. What makes one interpreter superior to another depends on how widely travelled the individual is and how many different types of landscapes they have been exposed to in addition to the formal interpreter training received. Differing geographic areas and their ability to blend real world knowledge and scientific procedures helps an interpreter come to the correct and final conclusion of any landscape assessment (Jensen 2005). A good interpreter can also see stereoscopically. This is the ability to use a stereoscope, view two adjacent photographs or satellite images, and manage to see the 3-dimensional effect without a problem. It is important for an interpreter to be able to delineate one feature, object or category of ground type from another which can be a challenge if any of those elements to not have a discrete edge but a fuzzy nearly imperceptible boundary. Thus a foundational knowledge of several classification systems is essential, such as knowing the characteristics of certain types of land use from land cover. What characterizes an area of "urban" versus "forest", "mangrove" versus "savanna", "pine forest" versus "deciduous forest"? In order to answer these questions two pieces of knowledge are essential: (1) being familiar with classification systems (there are many) that give specific criteria for different types of land use and land cover and (2) what represents each type of land use or land cover regarding distinguishing characteristic elements of shape, size, pattern, tone (or hue), shadows, texture, and others. Finally an interpreter must be aware of context to make the best assessment of any photo or satellite imagery. Contextual information could be the time of year, the scale, ancillary map data or on-the-ground field data to help corroborate (or refute) interpretation conclusions (Lillesand et al. 2008). It is useful here to elaborate on this point by showing the reader my own personal familiarity with these qualitative and quantitative techniques.

5.5.1 Personal Aside

From personal experience, my two longer field trips to Cuba, only a few weeks each, were at different times of the year, and each covered the opposite end of the island. On those two trips I brought my land cover and land use maps that I had created utilizing satellite image analysis, my field GPS, mapping and remote sensing software, plant identification books, and other land cover/use maps generated by other individuals and groups to compare to my own, as well as other reference materials. A map of those two field trips is shown below in Fig. 5.1 to illustrate the extent of these forays. Each dot is a GPS point, and associated with each point is a data record that chronicled the following data: latitude/longitude, 1–5 dominant land cover/use types within ~1–0.5 km radius from point (depending on topography of landscape), whether the landscape was heterogeneous or homogenous regarding land cover/use, elevation, and a description category. Interestingly, the description category was one of the most useful later on because it gave context to the rest of the information. A snapshot of one of these tables is shown in Fig. 5.2. It was from these

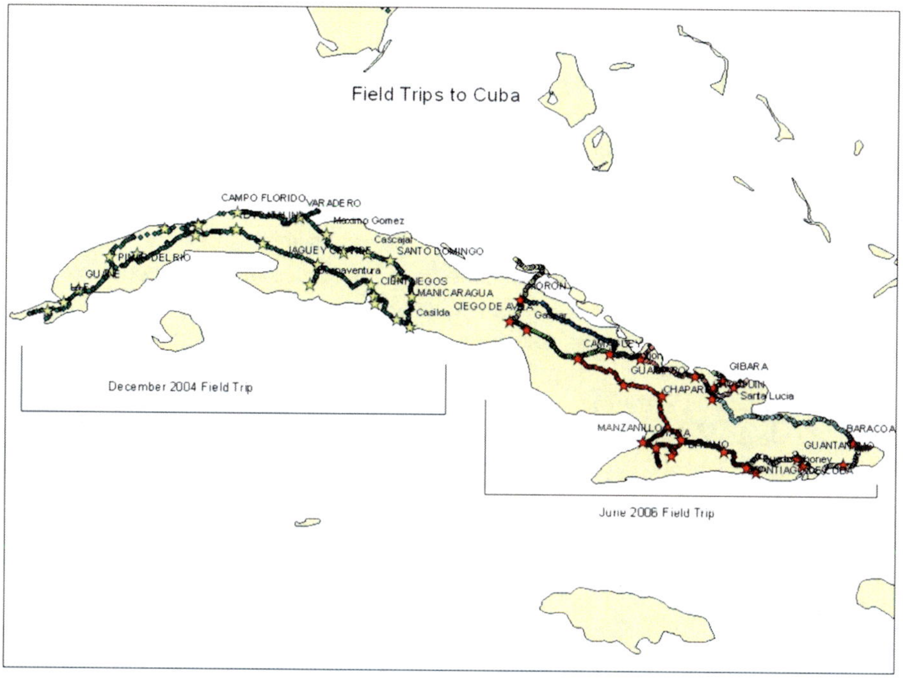

Fig. 5.1 Gebelein 2004 and 2006 field trips to Cuba

WAYPOINT	DATE/TIME	LATITUDE	LONGITUDE	ELEVATION	LANDSCAPE_	HETEROGENO	HOMOGENOUS	DESCRIPTIO
808	5/31/2006	20.014480	-75.782050	68	50% pasture	yes	no	pastures for horses, elders blooming, shrubl
809	5/31/2006	20.012710	-75.763840	221	80% pasture	yes	no	80% pasture mixed w/ poinciana, royal palms,
810	5/31/2006	20.008740	-75.749340	267	85% pasture, 15% trees	yes	no	pasture w/ palm, mango, small houses, shrub
811	5/31/2006	20.006100	-75.736080	375	80% urban	yes	no	urban w/ frangipani, royal palms, mango
812	5/31/2006	19.999910	-75.729250	393	50% pasture, 50% mixed	yes	no	pastures mixed w/ Mexican creeper, mango, po
813	5/31/2006	19.991960	-75.723240	342	35% pasture, 65% mixed	yes	no	pastures for cows, bougainvillea, poinciana,

Fig. 5.2 Subset of a data table from GPS ground-truth exercise in Cuba, 2006

data tables of on-the-ground observations that I compared classified satellite imagery of similar dates to assess the accuracy of how I had spectrally organized and mapped each land cover/use class. It was also extremely helpful that my entire PhD graduate research had been based on comparison of ground based observations versus satellite images; thus allowing me to draw from experience differences and similarities between the two information sources.

Thus to make the best landscape assessment possible requires a lot of different skill sets, analytical background and contextual information. Planning such field excursions anywhere is complicated, particularly so in Cuba. And it is a gamble whether certain pieces of field equipment might be taken from you at the airport when you arrive with a promise that you will get it back upon departure. This was the one reason I did not bring a spectrometer to Cuba on any field trips. Spectrometers are instruments which allow you to measure electromagnetic spectra at any given point on the land or coastal area. For comparative purposes it would have been ideal to take spectra at each GPS acquisition in addition to the other information gathered. However, these are extremely expensive instruments and I was advised by U.S., Mexican and Cuban colleagues not to bring this. Since 2006, however the price of field spectrometers has dropped significantly and it is likely I would take that risk in the future because the data would be extremely valuable. The complications of fieldwork in Cuba also pertain to access. For example, in all the parks and reserves I visited I was not allowed to wander freely but was accompanied by a guide at all times. This was ultimately a benefit because the guides were specially trained naturalists in that specific region of Cuba and their assistance in identification of plant species and history of their park was invaluable for contextual evaluation. However, many parks require you obtain a permit ahead of time (not stated on their website or if one places a phone call), some parks close without warning, and some are not accessible at all by car or foot. This was (almost) the case in one remote preserve we visited in the Sierra Maestra and we travelled 3 h to get there, only to find the bridge that led to the entrance was partially missing. After braving this challenge, we proudly stood at the entrance to the park, only to be told by locals that, because of the bridge failure, the park was closed until further notice. Additional challenges of fieldwork also presented themselves in terms of access to certain rural areas where it was unclear if the land belonged to anyone. It is always an interesting decision process to hike on unmarked land, or not. One of the texts I highly recommend, and consulted for my 2006 field work, is "Field Methods in Remote Sensing" by Roger McCoy. Lastly on this personal aside, the research labs and field stations in Cuba are numerous and located in every province (terrestrial and coastal), the scientists and researchers that I encountered in each and every one was highly qualified and very helpful. If you plan a field excursion to Cuba, do not limit yourself to those experts in Havana or another major city, but also look to the technical experts afield, they are an extremely valuable resource.

5.6 Digital Image Processing

Remote sensing data can be either analogue or digital. As discussed previously, analogue format includes hard copy photograph. The digital format is for example, Landsat satellite imagery which is made up of seven matrices of brightness values. It is these brightness values that we can use to categorize the different landscapes and land formations around us. There are other types of remote sensors that acquire data other than brightness values but in the interest of brevity there is a complete description of all types of remote sensors and different data types they acquire provided in Jensen (2005). The Landsat sensor gathers data about objects and landscapes in its field of view. The data that it records is the electromagnetic energy radiation (EMR) emitted, reflected or back-scattered from a geographic area or from objects of interest. The electromagnetic energy that is acquired in this process is utilized as a surrogate for the real landscape being researched; it is a data representation of reality in other words. This energy that is received by the sensor is what we analyze when performing digital image processing to turn that EMR into usable and useful information

(Jensen 2005). So now what is done with this acquired data? In layman's terms, digital image processing means that the image is fed into a software system that analyzes the image on a per pixel basis or on the basis of some grouping of pixels. These pixels are subject to various equations and the resulting statistics form the basis for a new digital image that can be shown either in a picture format or can be subject to additional processing steps to glean more information. Processing this data can be broken down into several steps. Image pre-processing is the first step, this corrects any geometric alignment problems, radiometric calibrations, and removes any noise from the image (haze in the air for example). Image enhancement is the next processing phase that includes techniques such contrast stretching, edge detection, texture separation, and intensity-hue colour space transformations. Image classification is the third step and the purpose of this is to use quantitative techniques to automate detection of the features of interest in the image. This involves taking the multispectral layers of imagery data and creating decision rules to determine the land cover or land use type of each pixel in the image. When this decision rule is used on the spectral information alone this is called spectral pattern recognition. And the image is "Classified" or categorized based on this set of rules and parameters. The image can also be classified using one of many spatial pattern recognition techniques by identifying shapes and linear features. In either case the result is categorizing the image to produce a thematic map of land cover or land use classes. From this point a researcher can then use these results to yield summary statistics of how much area is attributed to each class/category, and start to identify trends or patterns that repeat throughout the image for further refinement (Jensen 2005; Lillesand et al. 2008). One of the most powerful exercises that can be accomplished with this thematic mapping of satellite data is looking at how land cover/use categories have changed over time. For example in 1958 forests were 16% of Cuba's land, and in 1970 that number shrank to 15% but in the following three decades reforestation projects have increased that number to 20% of Cuba's landscape (Rodriguez 1999). This type of information is valuable as a physical representation of changing policies, habits and attitudes regarding the environment.

5.7 Geographic Information Systems (GIS) Overview

A Geographic Information System (GIS) is very useful in landscape analysis and characterization. Defining GIS can be challenging due to the fact that it is used in so many academic, research and business venues. The term has become very elastic; resulting in more and more confusing language as a result of an increased number of definitions becoming a part of both the scientific and popular literature (Demers 2000). Thus how it is defined here is specific to landscape assessment. For this purpose we can simply define Geographic Information Systems as:

> *An integrated system of hardware and software, methods and tools that allow for the processing of spatial data into information. More specifically, a software system that allows a user to create, acquire, manage, analyse, model, display and query spatially referenced data for the purpose of decision-making in reference to some portion of the earth.*

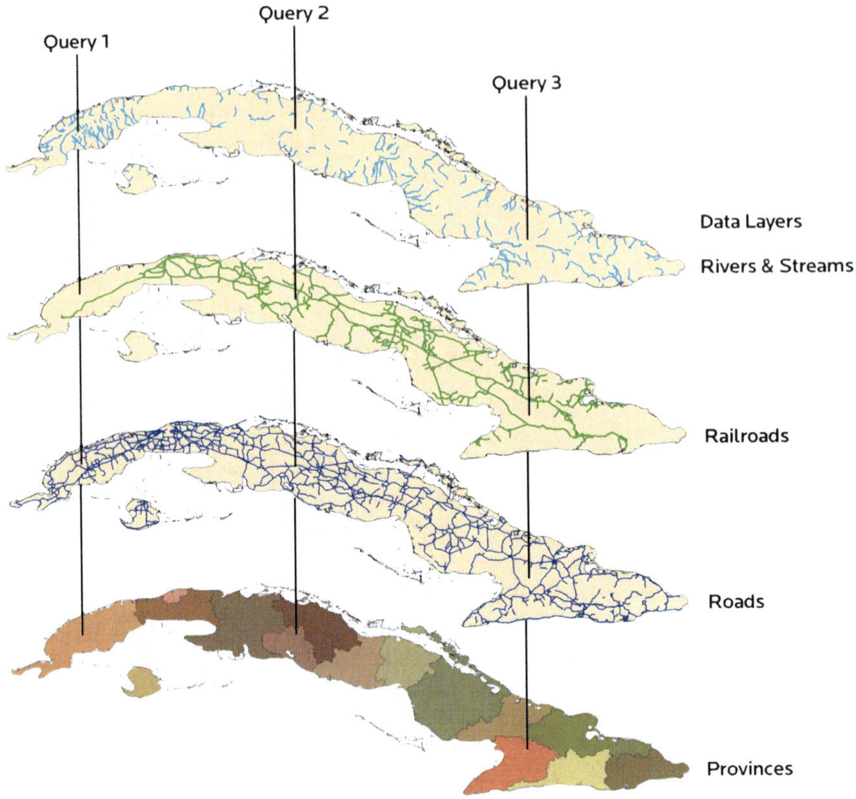

Fig. 5.3 GIS data layers illustrating same query/question, three different areas

The above definition is generally adapted from other authors of GIS textbooks, particularly Demers (2000), Clarke (2001), and Lo and Yeung (2002).

Given the above definition, what in fact, does a GIS allow a user to do with data? It allows the user to integrate many different types of data into one spatially defined area, ask a question of all those data layers to identify where commonalities/differences lie, or try and discover trends and patterns over this landscape, and the answer to that question: is given in a map. So while the integrated data that is *embedded in the map* is in tabular format (like an excel spreadsheet for example), plus a satellite image, plus a scanned aerial photograph, plus a surveyors map giving detailed contours of a complex landscape. The final map acts as an answer to a question that incorporates all of those pieces of data. No longer are the data confusing, but intellectually accessible to all who view the resulting map. This is especially important in research that describes landscape change over time simply because to truly represent a landscape in map format is extremely complex. To incorporate all the data necessary to make management decisions, plan, conserve or study an area requires that a person be extremely organized with how the landscape data is presented. Figure 5.3 illustrates sample GIS data layers of Cuba. The three lines indicate one

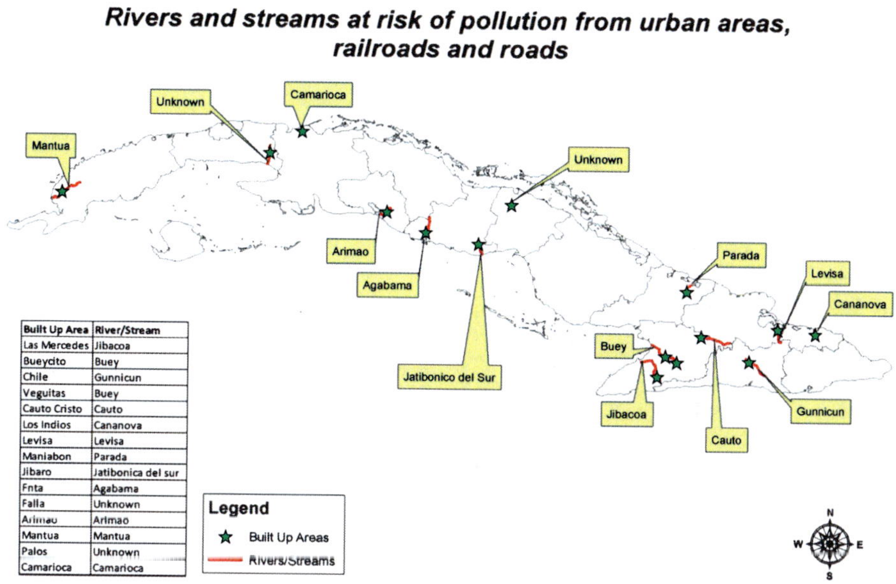

Fig. 5.4 Showing the GIS data layers, asking the data a question and the final map example

question that could be asked which uses data from all four data layers in three separate spatial locations. For example, say we are concerned because it appears that whenever there is a river or stream close to (or intersects) a railroad line or road, the amount of trash and sediment flow into the stream appears to contaminate it. Thus, we can ask the question: Where are the rivers and streams that intersect a railroad or road? The resulting map shows that out of 329 rivers and streams in Cuba, 187 are at risk. We can further refine this question by asking "Which of these 187 rivers or streams are within 0.5 km of a city or town?" The results show that of the 187 streams and rivers selected previously, only 14 satisfy this query (Fig. 5.4). The original data table that makes up the rivers and streams by itself has 329 individual records that are defined as either a stream or river, so clearly using a GIS to narrow down and identify which individual streams or rivers are at risk and illustrating the answer to that query in this fashion, is effective.

5.8 GIS in Cuba

There are several governmental organizations that do GIS mapping and analysis in Cuba. The first and probably most famous is the Institute of Geography in the Academy of Sciences in Cuba (Instituto de Geografia de la Academia de Ciencias de Cuba). This institute was founded in 1962 and works closely with the University of Havana's Department of Geography. The objective of this institute was initially to compile all geographic material, catalogue Cuban nomenclature, and create maps which ultimately became the foundation for the National Atlas of Cuba series. In 1989 the (most recent) National Atlas of Cuba was published and contains over 600 maps of population, agriculture, technology, education, forest, conservation areas, geology, soils and other topics that covered 24 different themes. Within this institute there are many departments, including several of interest here such as is the Department of Cartographic Modelling and Geographic Information. This department focuses on remote sensing and GIS modelling interactions between nature and society. Additionally, there is also the Department of the Environment, which focuses on deterioration of resources and how they are linked to economic and social problems. There is also the National System of Protected Areas (SNAP) which is within the Ministry of Science Technology and Environment (CITMA). Its mission is to guarantee conservation of natural resources and encourage sustainable development for Cuba. There are numerous research labs, university scientists and field biologists who also utilize GIS but they are too numerous to list here. An excellent online reference for these groups is *Redcien: Cuban Science Network*. This lists out every single category of science and research that is current and describes plans for the future. It also lists and describes the various channels through which this information is finally disseminated.

5.9 Mapping Agricultural Favorability and Productivity Using GIS

Cuba has experienced significant changes in its agricultural distribution across the island since the time of Columbus. Up until satellite imagery and GIS data became available for this type of analysis we can only try to estimate, from historical records and surveys, how agriculture replaced different natural land cover over time, and where this occurred. These changes to the landscape are documented in landowners and military surveys of the various types of crops important during each time period. As discussed in Chap. 1, there were dominant types of agriculture throughout the last several hundred years but clearly, the most important and detrimental to the natural land cover has been sugarcane. Sugarcane impacted the entire island either by being grown, or where timber was logged for fuel.

One of the factors that is always considered for establishment of any type of farming is the potential of the land for agricultural production. Digital data from the New Atlas of Cuba was used to generate this map (Fig. 5.5) using GIS to illustrate where the most and least favourable lands currently exist for agriculture.

This map is quite generalized but gives the reader a better understanding of the spatial layout of areas more favourable or less favourable for cultivation. The legend categories were derived from knowledge of the following: nutrient rich-poor soils, elevation, erosion potential, protected areas (terrestrial and urban), urbanized areas, precipitation, and areas currently occupied with some other activity such as grazing lands, for example. The two categories for "slightly favourable" refer to (1) That

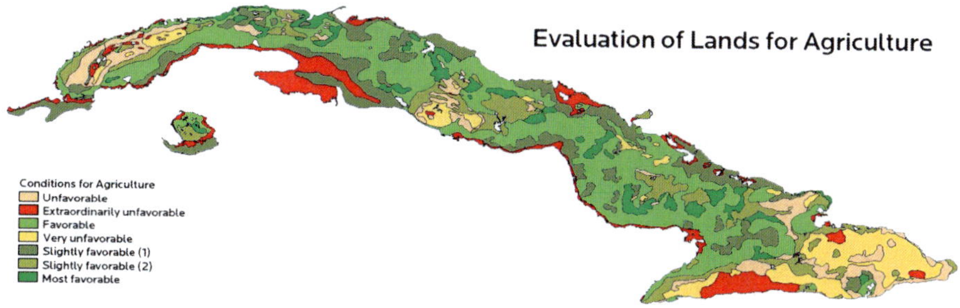

Fig. 5.5 Map of the best to worst locations for various types of agriculture (*Data from New Atlas of Cuba* 1989 *via Mapoteca Digital*)

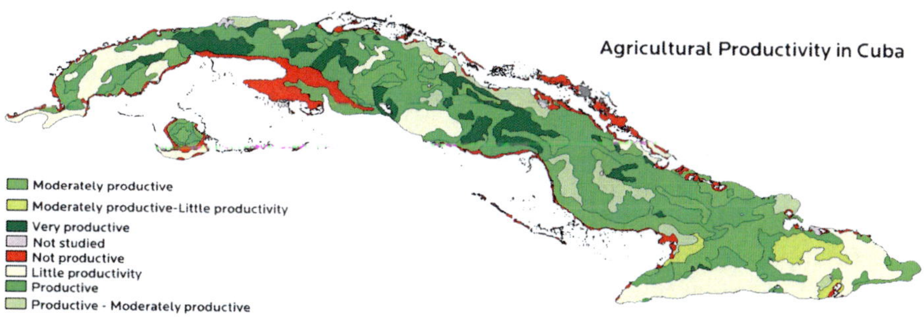

Fig. 5.6 GIS map illustrating agricultural productivity in Cuba

erosion can be a slight issue, drainage may be needed and (2) erosion is a moderate issue, working mechanical machinery is difficult here, watering the fields can be complicated.

We can compare the map shown in Fig. 5.5 to the Agricultural Productivity Map in Fig. 5.6 that illustrates overall farming productivity for all crops during the same time. Figure 5.6 was also generated from online GIS data from Nuevo Atlas Nacional de Cuba (1989).

5.10 Impacts of Urban Expansion and Development Using Satellite Imagery and GIS Data

The extent to which villages have expanded into towns, become small cities and then sprawling urban and suburban areas is an interesting process anywhere in the world. From early explorer accounts and city surveyors it is possible to take those original estimates of urban land use and depict them here. Clearly estimation from aerial or orbital imagery of such urban growth did not begin until early in the

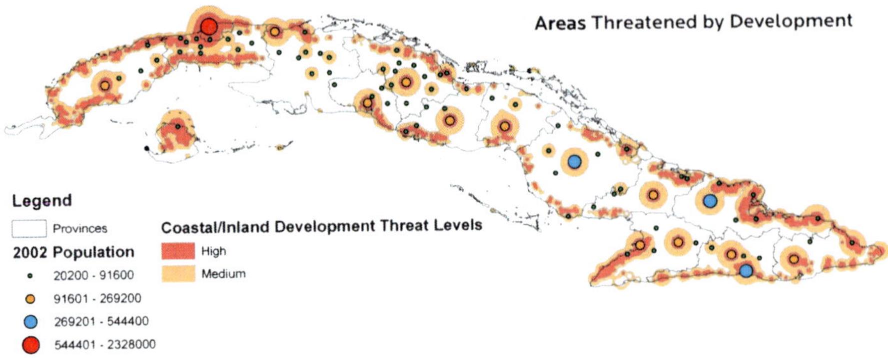

Fig. 5.7 GIS map showing point locations of population ranges associated with areas threatened by development and expansion as highly threatened or a medium level threat of permanent damage

twentieth century. Even so we can not only establish where urban areas currently exist, but also analyse the impacts they can have on the surrounding environment. There have been several studies which have generated excellent GIS data for this express purpose such as the World Resources Institute (WRI) and more specifically, the sub-group within WRI called Reefs at Risk. They conduct analyses and generate accessible GIS data for researchers, scientists and students to download and examine to discover trends, land use patterns, and model possibilities towards future environmental scenarios. The Reefs at Risk scientists and researchers compile data from numerous sources on the physical environments, coral reefs and their conditions, social and economic factors, pollution and other observed threats and physical influences.

For example in the following map the variables for Cuban population data from 2002 is combined with sewage discharge, urban runoff, tourist development and construction projects to model where development are a high or medium threat to the environment. Along the coastal areas reefs are the targeted threat area although clearly other aspects of the marine environment are threatened by similar factors (Fig. 5.7).

To take us further with this illustration of anthropogenic impacts on the Cuban environment is the following GIS map which clearly demonstrates the potential for sediment transport from inland areas out to the coastal zones, mangrove areas and other sensitive habitats (Fig. 5.8).

Sediment deposition is a very significant contributor to coastal degradation. Generally it is preceded by forest clearing and repeated removal of vegetation. If this removal occurs on semi-steep and steep slopes the deposition of sediment to the surrounding areas is higher because gravity pulls the water and soil towards lower elevations when a heavy rainfall event transpires. However, this does not have to take place on a steep slope to damage surrounding habitats. Agriculture is a major contributor due to the clearing of the land for the planting of crops. So, although it

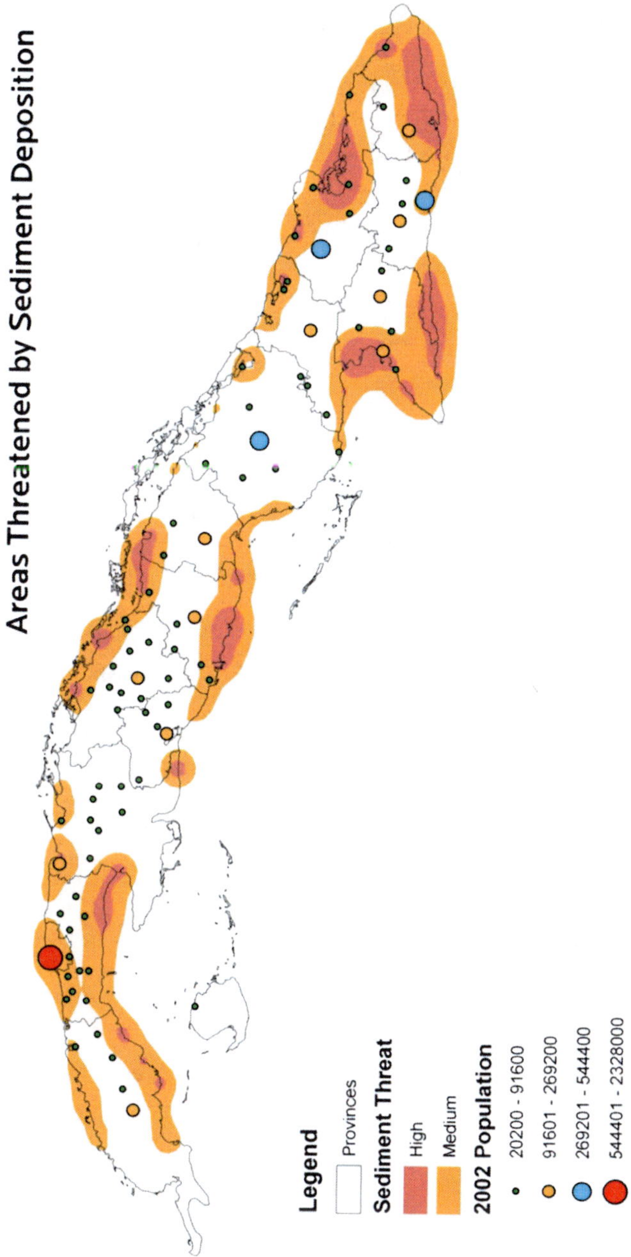

Fig. 5.8 This map shows us the areas that are threatened by sediment deposition in Cuba

is essential for its economic contributions and Cuba's food security, it is an important cause of nutrient runoff into sensitive shoreline habitats (Burke and Maidens 2004).

It is clear that technological developments are essential to identify patterns and trends in landscape changes both natural and anthropogenic. To have the ability to combine simple to complex sets of data and information, both qualitative and quantitative, is essential because the end product, a map, is a visualization tool that students, policy-makers, government officials, and scientists can understand and be the basis from which decisions are made on common ground. GIS and remote sensing are key to organizing data and infrastructural information into a common geographic space for issues of conservation both terrestrial, coastal and marine.

References

Altshuler J (1982) Space activities in Cuba. In: Fundamental aspects of materials science in space: proceedings of symposium 9 of the COSPAR twenty-fourth plenary meeting, Issues 5–8, Ottawa, 16th May–2nd June 1982

Antrop M (2005) Why landscapes of the past are important for the future. Landsc Urban Plan 70:21–34. doi:10.1016/j.landurbplan.2003.10.002

Arnaldo Tamayo Méndez (2011) In: Encyclopædia Britannica. Retrieved from. http://www.britannica.com/EBchecked/topic/1509108/Arnaldo-Tamayo-Mendez

Burke L, Maidens J (2004) Reefs at risk in the Caribbean. World Resources Institute, Washington, DC

Burrows WE (1998) This new ocean. Random House, New York

Clarke K (2001) Getting started with geographic information system, 3rd edn. Prentice Hall, Upper Saddle River

Day DA (1998) In: Day DA, Logsdon JM, Latell B (eds) Eye in the sky – the story of the CORONA spy satellites, Smithsonian history of aviation series. Smithsonian Institution Press, Washington

Demers M (2000) Fundamentals of geographic information systems, 2nd edn. Wiley, New York, p 7

Evans C, Lulla P, Dessinov L, Glazovskiy N, Kasimov N, Knizhnikov Y (2000) Shuttle-Mir earth science investigations: studying dynamic earth environments from the Mir space station. In: Lulla K, Dessinov L (eds) Dynamic earth environments. Wiley, New York

Fischer WA (1975) History of remote sensing. In: Reeves RG (ed) Manual of remote sensing. American Society of Photogrammetry, Virginia

Gebelein J (2003) Astronaut-acquired photography of earth: its history and continued applicability in quantitative analyses. J Space Commun 3: 1–25. http://satjournal.tcom.ohiou.edu/pdf/gebelein.pdf

Glazovskiy NF, Dessinov LV (2000) Russian visual observations of Earth: historical perspective. In: Lulla K, Dessinov L (eds) Dynamic earth environments. New York: John Wiley & Sons, Inc., pp 15–23

Jensen JR (2000) Remote sensing of the environment – an earth resource perspective. Prentice Hall, Upper Saddle River

Jensen JR (2005) Introductory digital image processing – a remote sensing perspective, 3rd edn. Prentice Hall, Upper Saddle River

Lillesand T, Kiefer R, Chipman J (2008) Remote sensing and image interpretation, 6th edn. Wiley, Hoboken

Lo C, Yeung A (2002) Concepts and techniques of geographic information systems. Prentice Hall, Upper Saddle River

Philipson W (1997) Manual of photographic interpretation, 2nd edn. American Society for Photogrammetry & Remote Sensing, Maryland

RedCien – Mapoteco Digital (1989) Nuevo Atlas Nacional de Cuba. http://mapas.redciencia.cu:8081/mapoteca/. Accessed 2010–2011

Rich BR, Janos L (1994) Skunk works – a personal memoir of my years at Lockheed. Little Brown, New York

Rodríguez E (1999) Cuban forests their history and characteristics. Instituto Cubano del Libro, Havana

Scambos T, Sergienko O, Sargent A, MacAyeal D, Fastook J (2005) ICESat profiles of tabular iceberg margins and iceberg break-up at low latitudes. Geophys Res Lett 23. doi:10.1029/2005GL023802

Thomas H (1998) Cuba or the pursuit of freedom. Da Capo Press, New York

Chapter 6
Cuban Environment, the Past, Present and Future

Abstract This final chapter concludes with a brief look at some of the most influential and recent factors impacting Cuba's lands and coastal waters in addition to the actors who contribute to their analysis and ongoing research for scientific and other purposes. Cuba's plans to conserve or develop are also briefly discussed in conjunction with the positive and negative activities impacting land cover and land use futures in Cuba with the influx of potential tourism.

J. Gebelein, *A Geographic Perspective of Cuban Landscapes*, Landscape Series 15,
DOI 10.1007/978-94-007-2406-8_6,

Keywords Green Revolution • Special Period • Reanimation of the Economy • Climate change • Biodiversity • Fragmentation edge effect • Scientist • National Center for Scientific Research (CENIC) • Environment Impact Assessment (EIA)

6.1 Introduction

The island of Cuba has experienced dramatic changes to its landscape since the time of Columbus. There has been an extremely high loss of habitat, particularly forests, and thus a decrease in suitable environments for the unique species of flora and fauna that survive today. There have been increasing demands on Cuba's natural resources including land for sugarcane, urban expansion, livestock grazing, mining and logging. Despite this Cuba currently has one of the highest levels of biodiversity in the Caribbean. In this chapter the factors which most influenced landscape change are summarized. Additionally, the most significant factors that impact coastal waters are also illustrated. It is very important to give credit and describe the efforts of the many exemplary scientists, conservation managers, other individuals and groups who have invested time, energy and for some, a life's worth of research protecting Cuba's natural resources and attempting to prevent the exploitation that occurred in the past, from happening in the future. Dovetailing this topic the current plans for development versus protecting certain areas is also discussed. Clearly while there is a need for various types of infrastructure and expansion in Cuba to support its residents, government and economy there is also a desire to protect the unique features, flora and fauna that distinguish Cuba from all other islands in the Caribbean. This attitude of protection and pride is one that I found consistent through all my travels and informal interviews with residents of Cuba. There might be differing, debatable issues when talking about politics. However when the topic of Cuba's distinctive landscapes and animal life arose during conversations, the swelling of the chest followed by the inevitable enthusiastic smile was always present and their pride visibly shone through. Finally, regardless of this desire to conserve and protect there are currently clear positive and negative elements influencing natural land cover. Cuba has many types of habitats that, if left unprotected, will be irretrievably damaged. However, there have been excellent laws and policies enacted which began the process of protection via the national park system and establishment of world heritage sites, for example. The future holds many uncertainties for Cuba in terms of its relationship with the United States, and possibility for sudden and greatly increased pressure from huge floods of arriving tourists to a place that simply does not have the infrastructure to support them. This is but one example of a future scenario that is discussed in this chapter.

6.2 Influential Factors on Cuba's Lands and Coastal Waters

6.2.1 On Land

The previous chapters have described and discussed many significant influential factors on Cuba's terrestrial lands, coastal areas and water. Much of the distant past has been discussed in previous chapters as well; thus here we conclude with the current and future of Cuba's lands and coastal waters. One of the major foci that should stand out to the reader is that the changes to the land and seascapes were a reflection of the needs of the people, ruling entity, economic requirements of the period, cultural or political drivers and Cuba's relationship to the rest of the Caribbean. As discussed in previous chapters one of the most significant categories of impact over the entire history of Cuba is the influence of agricultural practices. Before 1989, agriculture in Cuba could be characterized as significantly monoculture, with large scale irrigation, extensive mechanization, and elevated quantities of industrial inputs (Funes 2002). This agricultural industrialization was extremely reliant on the close economic ties to the Soviet Union and Soviet Bloc countries. Therefore with the eventual collapse of the Soviet Union, those ties disappeared along with all of their associated economic and industrial support for agricultural production in Cuba (Nelson et al. 2009). Regarding Cuba's state versus non-state actors in the agricultural sector there are some key players to note. Recently, several of the most well-known non-state actors in the Cuban agroecology movement are the Cuban Association of Agriculture and Forestry Workers (ACTAF) and the National Association of Small Farmers (ANAP). These non-state actors focus their energy on sustainable agricultural methods as well as education and outreach to farmers. The Cuban state's role has been to establish policies that successfully regulate production decisions accessible to individual farmers and thus force agroecological farming techniques (Nelson et al. 2009).

Research conducted by both Cuban and international scholars has provided data and analysis that has narrowed the focus of recent agricultural transformations into three periods of time, the Green Revolution, Special Period and the Reanimation of the Economy. Because of their high level of influences on Cuban landscape alteration and ultimately, its future, we will begin with these factors first.

6.2.2 Green Revolution

The Green Revolution occurred from 1984 to 1991 and was carried out with the assistance from the Soviet Union so thus is also known as the "Soviet Agricultural Revolution" (Warwick 2000, 2001). The goal of this was more on increasing agricultural output than ecological preservation. The focus of this period was to boost productivity by allowing farmers to establish crops on lands that had been previously protected. This meant that there was substantial clearing of forests and other native vegetation communities that had previously been left undisturbed. The agricultural focus was mostly on sugarcane cultivation, which compounded the problems associated with monoculture cultivation of sugarcane for export. With this "green revolution" was an increased dependence on imported pesticides, use of improved hybrid seeds, and more dependence on the Soviet Union for machinery and oil (Rosset 1997; Warwick 2001). During this time period the environment that had been cleared and utilized in this manner was seriously negatively impacted as described by Maal-Bared (2006) and Funes (2007) (Febles-González et al. 2011).

6.2.3 The Special Period

The Special Period had very distinct sets of environmental issues before, during and after that time. Before the Special Period the threats to the environment were freshwater quality issues, the use of high amounts of pesticides which led to issues of groundwater and other pollution issues, intensive strip mining activities, industrial wastewater and sludge, terrestrial soil deterioration, freshwater quality issues and surface water stressors. These activities were proceeding with minimal active protections for these increasingly threatened habitats and their associated flora and fauna. The delicate ecosystems of the bays, inlets and many other types of coastal areas had become very polluted from these types of stressors. In contrast to this, the fall of the Soviet Union and collapse of trade relations with Soviet bloc countries had unforeseen benefits for preservation of these areas, and associated increased pressures on others. There was a reduction of dumping waste into rivers and bays, a decrease in gaseous discharge from industrial operations into the atmosphere, and diminishing exploitation of inland and coastal water resources (CITMA 1992). However, despite the lessening pressure from certain industries on the Cuban landscapes, this was not all encompassing. For example because of the sudden lack of access to gas and other fuel resources, deforestation increased during this time period in some of the most environmentally delicate areas such as the Zapata preserve and mangrove forests nearby coastal cities and towns. Curiously, there were some factors which, although severely impacted by the resulting Special Period, changed their method of impact during this time. So, while the prior techniques of pollution were greatly reduced or gone, they were replaced by different modes of pollution. They are agriculture pressures, deforestation, dam construction, tourism, military operations and certain types of mining (Maal-Bared 2006). These issues related to the Special Period are discussed more in depth in Chap. 2. Another important influential factor related to this is the outcome and long term effects of the policies that resulted from the Special Period. Of particular note are the greater efforts to maximize organic farming techniques such and IPM (Integrated Pest Management), careful crop rotation to maximize soil nutrient retention and minimize weed invasion of an area. This is especially important to the future of Cuba's land and coastal waters because it minimizes soil erosion. In previous decades and centuries of historical land use, no regard was given to removing the natural vegetation, or getting multiple harvests from an area with no thought to moderate nutrient depletion. Thus when the land no longer offered the crop yield, farmers simply moved onto another plot of land to repeat this process. The result of this had been increased erosion and soils being washed out to coastal areas, causing accumulated sediment issues there, or being transported to lower elevations and changing flora composition there over time. This perception has dramatically changed since the Special Period both out of necessity but also from an increased awareness of natural habitat destruction and what it means to reduce the biodiversity in Cuba (Maal-Bared 2006).

6.2.4 *Reanimation of the Economy*

This period of time overlapped with the Special Period and began in the late 1990s lasting until 2007. Some would argue it is still in progress in 2010–2011. What marks this period is the successful transition from dependence on chemical intermediates to biological agents in a agro-ecological conversion process that scientists have taken advantage of for larger scale farming systems (Funes 2002; Febles-González et al. 2011). Despite the successful transition in organic nutrients and pesticide management strategies during this period, in 2007, only about two million hectares of land were actively cultivated, in spite of the fact there were over four million hectares of fertile lands available. During the early years of the Special Period there was a national effort to stimulate planting food. However stimulation did not last through the reanimation of the economy period and from 1999 to 2007 and cultivated land decreased by 17.5% overall. This is a significant change in land use and the reason behind it is that while Cubans may have access to the lands and the desire to farm them, they do not have the necessary resources to produce a crop yield. This decrease is not only in one area in Cuba, but exists in every province according to data analysed from 2002 to 2007 as shown in Fig. 6.1 below.

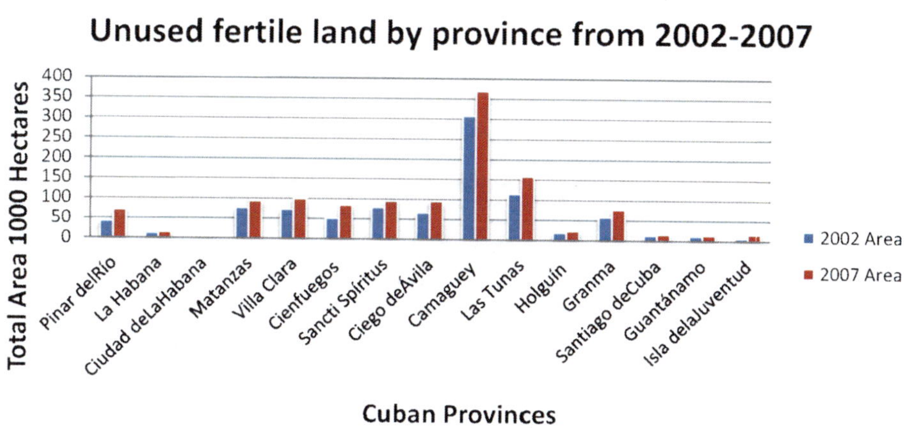

Fig. 6.1 These data clearly show the increase in unused fertile agricultural lands in every province comparing 2002–2007 (*data from* Febles-González et al. 2011)

This lack of resources creates serious problems for increasing yields. If the people are not given the tools they requires to harvest the lands, it is impossible for them to achieve goals set by the Reanimation set forth by the Cuban government (Febles-González et al. 2011).

The resulting land use changes that follow this sort of failure to cultivate fertile lands is the land slowly converts from bare soils to the various stages of successional vegetation that naturally occurs when fields are allowed to go fallow. Grasses and weeds quickly take over an area, followed by small woody shrubs and eventually larger shrubs and small trees. If left alone long enough the area (depending on soil type, elevation and distance from inland or coastal waters) will require a large amount of effort to return it to its former cultivatable structure. During my field research particularly in 2006 this pattern was widely evident in the many abandoned fields in the middle (from Holgiun) and southern (towards Las Tunas) portion of Cuba all the way to the Santiago de Cuba near the Sierra Maestra Mountains in the east.

From 2007 to present there has been a restructuring of the agricultural production programs to create new agro-ecological paradigms which will allows for higher productivity, environmentally and energy efficient methods, and biologically stable systems. This new paradigm is also focused on decreasing environmental pollution, protecting natural resources, and providing the Cuban population with sufficient food (Funes 2001). One example of this applied concept is shown by Febles-González et al. (2011) where, since 2001 organic manure enhanced by biofertilizers is being utilized to improve soil fertility and ultimately, reduce erosion.

6.2.5 *Climate Change and Indicator Species on Land*

Cuban conservation was explored in the recent study completed by the United Nations on the state of the environment in Latin America and the Caribbean. The information presented there showed several grim accounts of the current state of the environment such as the fact that the extraordinarily high level of biodiversity in the Caribbean is being lost or is gravely threatened by anthropogenic activities shown in Fig. 6.2 below.

Latin America and the Caribbean: Countries Among the Twenty with the Greatest Number of Endangered Plant and Animal Species

	Endangered Animal Species		Endangered Plant Species
Mexico	636	Brazil	382
Colombia	429	Peru	275
Ecuador	369	Mexico	231
Brazil	356	Colombia	223
Peru	261	Jamaica	209
		Panama	194
		Cuba	163

Source: IUCN, 2008a

Fig. 6.2 The top endangered plant and animal species, a comparative chart of the Caribbean case (UNEP 2010)

Unfortunately, biodiversity loss is an observable fact that is happening at all levels and throughout the Caribbean region. Many times the health of an ecosystem, habitat or biodiversity level can be linked to what are known as indicator species. One of the linkages between biodiversity loss and human impacts is the golden frog (*Atelopus zeteki*) which is on the brink of extinction in its natural habitats. This endangered species has suffered from an attack by the chytrid fungus (*Batrachochytrium dendrobatydis*), that has killed off several species of frogs and toads in very large numbers all over the world. The spread of this fungus is linked to increasing temperature related to global warming (Pounds and Coloma 2008) as well as large and small-scale weather fluctuations as additional influences on the high rate of extinction in that region. Global change is impacting species around the world but it is beginning to effectively reduce or extinguish species in the tropics; and not only at high latitudes where this type of impact was first noticed. This species loss, and thus biodiversity loss, is aggravated by a decline of overall area and fragmentation of habitats. The overall habitat availability has been decreasing on Cuba for quite some time, but there comes a point when species become extinct because there is simply nowhere else to go. Habitat fragmentation occurs when a species that required a certain amount of space for its territory can no longer expand into new territories for its male and female species to mate, nest, or forage. The habitat literally becomes so fragmented because of deforestation (for example) that there is no longer a continuous tract of land for a species to inhabit, and this puts an enormous amount of environmental stress onto any species of animal.

This is also true for plants as well. An area with a high level of flora biodiversity contains an entire community of plants that coexist, compete, and reproduce. This bio-complexity becomes threatened even when areas are set aside for protection. For example, in Cuba there are many cultivated areas that are (and have been) developed adjacent to protected areas. There has been, in the past, the assumption that setting aside an area for protection was good enough, with no real regard for the perimeter of the conservation area. This presents a problem. When transitioning from one vegetated area to another, there is a transition zone. There is almost never an abrupt change in a landscape from one land cover type to the next, just as there are almost never perfect geometric shapes in nature, there are transitions instead. For example, in the mountains of the Sierra Maestra there are wet forests of ferns, mosses, pines, and shade loving ground cover. The entire habitat there is always at least 50% indirect or dappled light, never full sunlight. If a large area is cleared and a preserve is created in the shape of a square, this can present a problem because now, there is direct sunlight streaming in all four sides of the habitat. This allows sun-loving vegetation in to colonize the soil floor and allows for more nutrient competition as more competing plants being to grow in these peripheral conditions along the edge. This is also known sometimes as a fragmentation edge effect, and can be devastating to plant communities. These type of pressures have been happening for a very long time in Cuba, however, now that there are so many endangered and threatened species of plants and some animals, it is becoming obvious that different protection measures must be taken to preserve the species that are left. In fact according to Giam et al. (2010) Cuba was grouped with only ten other countries in the world that were identified as countries with the greatest conservation needs, it was included in the top 20% for future plant species endangerment. This was done by relating the present level of plant species endangerment with (potential) total future habitat loss due to two factors: land use change and climate change (Giam et al. 2010).

There is another interesting trend occurring throughout Latin America which is that tree plantations have recently begun replacing primary forests, as well as being used in restoration programs and are being showcased as carbon sinks in mitigation programmes focused on climate change. Data from 2005, for example, shows that these tree plantations counted towards 4.2 million hectares in Cuba, Peru, Argentina, Mexico and Uruguay combined. These countries were in the minority considering that 40% of the total number of tree plantations in Latin America were found in Brazil (5.38 million hectares), and Chile with 2.66 million hectares. The reason for mentioning this here is that many countries throughout the Caribbean are reporting that their forest cover is rebounding from the drastic deforestation measures from their past. However even though these reforestation projects are good for reduction of soil loss and erosion, they are usually not counted as natural forest, they are counted as some other category of secondary or supplementary growth to the nation's vegetative land cover (UNEP 2010).

6.2.6 Waters (Influential Factors)

One of the greatest influential factors on water or land is pollution. Water pollution is a critical threat to Cuba's coastal ecosystems and can impact them in the form of inorganic or organic pollutants from industrial effluences, sewage or runoff. This type of nutrient overloading on a system can result in major fish kills, degraded seagrass communities, shellfish poisoning, algae blooms, and coral reef deterioration (NAS 2000). This has been a massive problem in the United States that now has a very large dead zone which is the result of nutrient pollution.

Another significant problem for Cuba is the treatment and disposal of wastewater. According to the Rio + 10 Report, greater than 70% of domestic wastewater, including a high percentage of the human sewage from Havana, is not treated at all or receives minimal treatment prior to getting discharged into bays, rivers and other coastal waters. The same report indicates there are 2,112 large water pollution sources in Cuba and less than 1 in 5 effectively treat their wastewater discharges. The three top polluter origins are human waste, discharges from food facilities, and waste matter from sugar refineries. There is also sediment and nutrient runoff from agricultural areas for example, effluent from sugar plantations, and pollutants from harbour activities also contribute to the pollution levels in Cuba's coastal areas (NAS 2000).

A third major issue are mining areas located on the coast such as Moa, where strip mines exist for, primarily, nickel and has destroyed the surrounding coastal habitats. This type of mining operation is still active and needs significant improvement if habitat restoration is possible in the many mining areas in Cuba inland and coastal.

The Cuba Subsystem of Marine Protected Areas (SAMP) supervises and protects any area that has a coastal or marine component to it including submerged coastal zone, offshore keys and coastal wetlands. The SAMP's evolution into a management entity is part of a collaborative effort among researchers in the Institute of Oceanography (IDO), the Ministry of Fishing Industry, the first MPA proposals considered by the National System of Protected Areas (SNAP), and the National Center for Protected Areas (CNAP) which initiated a planning process for the SAMP organizational structure and responsibilities. The SAMP has significant influence on the activities that occur in the submerged coastal zone, offshore keys and coastal wetlands including a strong voice in what is allowed and prohibited inside "public use zones" versus "zones of socioeconomic use". Thus the SAMP is one of the key organizations that works closely with other ministries, researchers and international groups to attain conservation goals set by the overarching national protected area plan (CNAP 2002).

6.3 Current Actors Impacting Cuba's Landscape and Coastal Waters

Cuba has a large number of technology specialists, researchers, scientists in all biological areas (marine and terrestrial), conservationists and naturalists in relation to its population (UNEP 2010). Since 1959 these centers of research, governmental organizations, conservation groups, park managers and scientific positions have all increased significantly. In terms of Cuba's visibility in international scientific publications, a study was done that showed a consistent output from 2001 to 2007 as shown in Fig. 6.3.

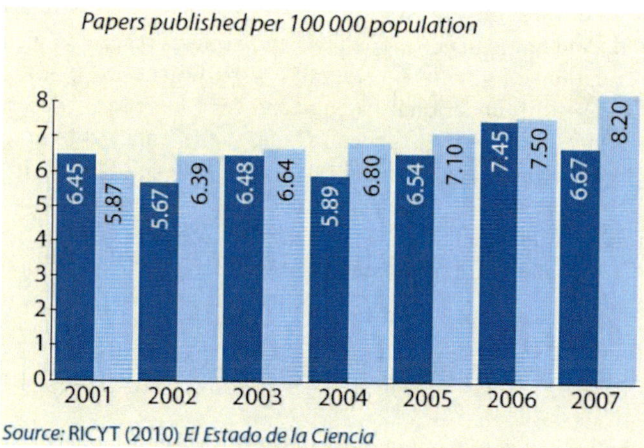

Fig. 6.3 Cuban visibility in scientific publications 2001–2007 (UNESCO Science Report 2010)

It is absolutely imperative that scientists continue to collaborate on the types of projects that generate peer-reviewed journal articles, conference proceedings, workshops and other types of partnerships. One of the main reasons for this elevated importance for Cuba is the government relies heavily on its scientists, researchers and their continued work on the state of Cuba's land use and land cover. The majority of the current research centers in Cuba began as research groups. Several of these centers, such as the National Center for Scientific Research (CENIC) performed a vital role in educating young science students and eventually in the establishment of multiple other research institutions. With this type of knowledge and collaboration Cuba can try to avoid the disaster that the influx of tourism has become on other islands in the Caribbean.

To dovetail from the above point, as previously discussed Cuba also has many official institutes, centres, and relatively new legal instruments that aid in the protection of the island's potentially exploitable resources. One of the most valuable new legal protective weapons is the Environment Impact Assessment (EIA). The EIA is the initial step in the process of granting or denying an environmental license. This was constructed under Law 81 and its underlying function is to make certain that decision makers have reviewed the proposal and thoroughly evaluated the possible environmental impacts of a proposed project prior to allowing that project to continue. The EIA requires that the project proposer detail how they plan to minimize, avoid and mitigate environmental impacts. The EIA is alerted to a potential project because any type of (large) construction effort that is proposed must have an application for an environmental license into the Centre for Inspection and Environmental Control (CICA). CICA resides within CITMA. EIA's have become an effective major stopping point for poorly planned projects. An EIA is obligatory for nearly all large construction projects. Additionally, EIA's are usually required for other building activities such as mining, roads and other transportation infrastructure, sewage storage, management and treatment structures, tourism infrastructure particularly those in coastal areas and all other activities that could forever change or damage a delicate ecosystem *or* somehow could block public access to natural resources in general (Whittle and Santos 2006). The laws and policies regarding the Cuban environment and relationships with development for tourism are many. Some fairly introductory but solid foundational background for this topic is described well in (Whittle and Santos 2006); 2003 Special Issue of the Tulane Environmental Law Journal titled "Environmental law and sustainable development in twenty first century Cuba"; 2004 CNAP document titled "The National System of Marine Protected Areas in Cuba", and the "National System of Protected Areas – Decree-Law No. 201 1999" in the Environmental Law in Cuba Series of publications.

6.4 Current Plans to Protect or Develop?

Cuba has recently begun expanding its search for oil and gas outwards into coastal areas and is working with foreign investors to conduct exploratory drilling off its northern coast, something that United States officials are increasingly uncomfortable

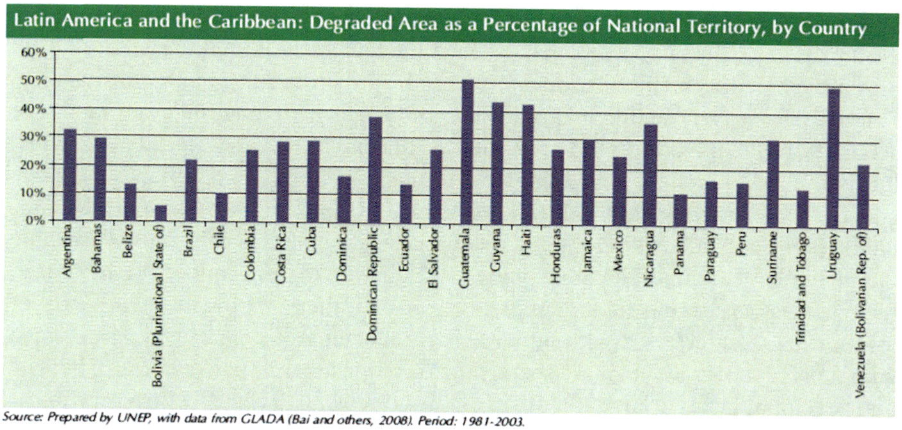

Fig. 6.4 Graph showing the degraded areas as a proportion of national territory (UN Report page 74, with permission) (Source: Prepared by UNEP, with data from GLADA. Period: 1981–2003)

with given the recent Deep Water Horizon drilling-associated spill in 2010. However there are untapped offshore reserves of oil and gas that would aid the struggling Cuban economy tremendously if allowed to be developed and exploited. In fact, Repsol announced in 2010 that it would proceed with exploratory drilling in Cuban territorial waters. If a disaster struck an oil platform off the northern coast of Cuba, the result could pollute vast areas of American coastline and devastate marine and coastal habitats. Since the United States and Cuba share this expanse of water between them, it would benefit both to begin a dialogue and establish protective measures against such a disaster and develop appropriate response structures (Pinon and Muse 2010).

Cuba also has yet undeveloped mineral resources on its terrestrial landscape, but whether Cuba will exploit those land-based and marine-based natural resources in a sustainable way remains to be seen. As the following concluding paragraphs will show, Cuba is more and more in the spotlight as an environment that could avoid some of the developmental land use change disasters that many other Caribbean nations have allowed over time.

In terms of other serious terrestrial issues, 14% of the world's land degradation occurs in Latin America and the Caribbean, and is continued development influences approximately 150 million of the region's inhabitants. It is obvious from this comparative graph (Fig. 6.4) provided by the GEF-UNEP-FAO project GLADA (Global Assessment of Land Degradation and Improvement), that the natural land degradation has continued as a general trend in Cuba showing that approximately 29% of Cuba's territory has deteriorated over time. Cuba is not the worst offender however; the most serious in Mesoamerica, where land degradation impacts 26% of the land, while in South America 14% of the land is affected. The highest levels of degraded lands are found in (51.3% of its national territory), followed closely by Uruguay (49.6%), Guyana (43.4%) and Haiti (42.6%) (UNEP 2010).

Future development in marine and terrestrial areas are currently threatened by an unparalleled upsurge of development targeted towards making Cuba the top tourist destination in the Caribbean. Overall tourism is increasing at an average rate of 10% per year and new large hotel complexes are being built at a rate never seen before. To give the reader an idea of the unprecedented speed with which tourism development is occurring; numerous new roads and bridges are being constructed each year that link the mainland to offshore keys, almost 3,000 new hotel rooms are added every year, once sparsely populated areas are now being inundated with commercial development which then require supporting infrastructure such as wastewater treatment plants to sustain them. At the moment Cuba has joint business enterprises with hundreds of financier from more than 35 countries who represent over 30 business sectors. Most of these foreign investors are backing the development of tourist facilities such as new hotels and supporting tourist infrastructure. The Cuban government has worked extremely hard to rebuild its economy and with the relatively new Ministry for the Environment, establishment of official policies and creation of protective designations for various types of terrestrial and marine habitats. The environmental agencies, institutes and centres, have fairly little funding compared to similar organizations in the United States and other countries in the Caribbean, yet the commitment of the individuals staffing those agencies is commendable considering the resources they *do* have and the continued commitment to keeping environmental protection at the top of the national, regional and local agenda while keeping the development rate relatively low (Whittle and Lindeman 2004).

6.5 Cuban Attitude Towards Environmental Conservation Issues?

6.5.1 Reflections on Preservation

The Cuban attitude towards environmental conservation issues is interesting. Cuba has a 95% literacy rate and its people are considered extremely well-educated. The United Nations involves many scientists and researchers from Latin America and Cuba is no exception. Therefore the dissemination of information might be slightly slower due to lack of televisions in every household, for example, but access to internet is becoming more common in larger cities and *is* common in all universities. There are also many radio broadcasts in Cuba which describe the latest contributions by Cubans to world affairs and important scientific research. And because of the huge number of Cubans who work for environmental agencies, field stations, ministries, labs, universities in flora or fauna research and are in every province, the education of the common Cuban individual about the importance of environmental conservation is very high. There are educational programs in every province that involve school children from a very young age to conduct different types of biological field work and come into regular contact with adults in their local community who are actively engaged in the preservation of the island's heritage. In fact CITMA is very engaged with the Ministry of Education in Cuba, thus the interrelationships among the various agencies throughout Cuba makes it possible for children to appreciate and understand cause and effect regarding pollution, clean drinking water, healthy reefs, and forest pressures.

6.6 Summary of Positive and Negative Elements Influencing Natural Land Cover and Future Possibilities

The previous chapters have shown that Cuba's landscapes have changed considerably from the lush forests and savannas seen by Columbus. From the first Spanish explorers to the current ruling party there have been some profound shifts in attitude or perception of Cuba's landscapes. The evolution of attitude toward the environment, perceptions of conservation, exploitation of forests by logging for fuel and harvesting of precious woods, slashing and burning large tracts of grasslands, savannas and forests for agricultural purposes, urban expansion and impacts of a growing population on the lands, consumption of huge areas of land for sugarcane and other crops and their associated infrastructure for export, and actively destructive activities linked to economic growth such as mining all have contributed to the negative impacts on Cuba's natural land cover. However, those negative elements that have so damaged the natural land cover of Cuba have also led to a greater appreciation and active protections in the forms of laws, policies and active individuals and research groups who are part of the effort to protect Cuba's heritage from disappearing forever.

Cuba is at a critical juncture. With the collapse of the Soviet Union just over 20 years ago and the related loss of billions in financial assistance followed by a long time of crisis for Cuba and its environment forced the government to reconsider every facet of its economic future. The baseline for Cuba's economy crashed when the prices of sugar fell following the Soviet bloc's overall withdrawal from trade with Cuba, and the trade agreements Cuba had with the Soviet Union evaporated in a very short period of time leaving Cuba and its people without basic requirements for consumer goods, fuel or materials to support industrial processes throughout the country. Since this time Cuba has slowly adjusted to this situation and what has emerged is a delicate and continually evolving relationship between economic growth for Cuba's government and its people and conservation measures. In the past these two facets that determine Cuba's landscape health have been at odds and the environment has usually been the loser. Interestingly, as global environmental awareness has increased, Cuba has had to respond in kind to guarantee more access to international markets by meeting specific international environmental quality targets, make itself subject to more voluntary regulations and open its environmental governance to other nation's certification standards and systems

(UNEP 2010). In response to this Cuba has indeed involved itself in more environmentally focused international symposiums, conferences, workshops, and scientific collaborations.

In 2002, at the World Summit on Sustainable Development in South Africa, the Cuban Ministry of Science, Technology and the Environment (CITMA) gave its Rio + 10 Report describing what it judged to be the Cuba's foremost environmental problems. They listed the following issues: loss of biological diversity, human settlement deterioration, inland and sea water pollution, soil degradation and deforestation (Whittle et al. 2003). These priorities have not changed much because even though there was a change of focus from issues before and then after the Special Period, and as Cuba's economy has *begun* to recover with more foreign investments we have seen a reversion back to some of the industrial problems that were present several decades ago.

In June 2004, the Caribbean Marine Biodiversity Workshop was held in Isla de Margarita, Venezuela. Cuban and other representatives from Caribbean nations gathered to discuss current conservation organizations and their research foci as well as discuss development possibilities with oil companies interested in investing in the Caribbean region. This workshop is another excellent example of a setting in which strong relationships were forged among research groups with similar interests in marine preservation such as the Smithsonian Research Institute (STRI), the Harte Research Institute for Gulf of Mexico Studies (HRI-Biodiversity of the Gulf of Mexico Project), FishBase, the Caribbean Coastal and Marine Productivity (CARICOMP), the Nature Conservancy, and oil companies such as Chevron, and ConocoPhillips (Miloslavich and Klein 2005). One of the goals of Cuba was to continue and strengthen their relationships with international research groups to ultimately determine the total biodiversity of the Gulf of Mexico. This conference yielded an excellent reference to this topic titled "Marine Biodiversity of the Cuban Archipelago: an overview" (2005) which describes 30 years of their own research data as well as a compilation of older marine-based observations since the 1600s. It is an interesting comparison to view Cuba's hundreds of miles of untouched coastline and compare the devastation that has taken place with regard to tourism development and population growth in countries like the Dominican Republic.

A positive outcome of Cuba's relative isolation with regard to development is that long stretches of coastline are still undeveloped and in these shallow waters adjacent to many areas in Cuba the marine diversity is unmatched in the Caribbean (Whittle and Santos 2006). It requires substantial monetary investment and many approval processes to even begin construction on a coastal area. However, this type of investment could potentially come from investors in the U.S. should the embargo be lifted.

One of the key scenarios to consider is if or when the U.S. Embargo lifts or is eased how will Cuba cope with the sudden influx of an additional 5–10 million tourists per year? This is likely a very low estimate given that for the 2010 calendar year, 82.6 million people visited Florida (VISIT FLORIDA(R) Research 2010). Granted, only 42.8 million visited by air, and the rest drove or took other means of transportation. By comparison the visitors to Cuba have increased from 1.7 million tourists who visited Cuba in 2001 to 2.53 million in 2010 (U.S. State Department 2010). That is a difference of 40.27 million people *per year*, using the lower Florida estimate.

Opening this tourism door would create sudden challenges for the current environmental and fiscal policies. An influx of 5–10 million people annually would require an enormous economic investment in roadways, sewage treatment facilities, hotel and resort construction, energy reliability in remote areas, and availability of popular consumer goods and services such as golf courses or other resort related activities. The success or failure of how well Cuba adapts to such an onslaught of people and pressures those numbers imply will rest on how well the recently established environmental organizations follow through on the laws and policies in the face of this incredible demand on the landscapes of Cuba.

References

Centro Nacional de Areas Protegidas (CNAP) (2002) Sistema Nacional de Áreas Protegidas, Cuba. Plan 2003–2008. CNAP. 202 pp

CITMA (Ministry of Science Environment and Technology) (1992) Referring to the negative repercussions of the blockade in the development and environment field. Cuba' Academia de Ciencias de Cuba, La Habana

Febles-González JM et al (2011) Cuban agricultural policy in the last 25 years. From conventional to organic agriculture. Land Use Policy. doi:10.1016/j.landusepol.2010.12.008

Funes F (2001) El movimie nto cubano de agricultura orgánica. In: Transformando el campo cubano. Avances de la agricultura sostenible. ACTAF—Food First—CEAS, La Habana

Funes F (2002) The organic farming movement in Cuba. In: Funes F, Garcia L, Bourque M, Perez N, Rosset P (eds) Sustainable agriculture and resistance: transforming food production in Cuba. Food First Books, Oakland, pp 1–26

Funes F (2007) Agroecología, agricultura orgánica y sostenibilidad. Asociación Cubana de Técnicos Agrícolas y Forestales, La Habana

Giam X et al (2010) Future habitat loss and the conservation of plant biodiversity. Biol Conserv 143:1594–1602

Maal-Bared R (2006) Comparing environmental issues in Cuba before and after the special period: balancing sustainable development and survival. Environ Int 32:349–358

Miloslavich P, Klein E (eds) (2005) Caribbean marine biodiversity, the known and the unknown. DEStech Publications, Inc, Lancaster

National Academy of Science (2000) Clean coastal waters: understanding and reducing the effects of nutrient pollution 1. National Academies Press, Washington

Nelson E, Scott S, Cukier J, Leyva Galab A (2009) Institutionalizing agroecology: successes and challenges in Cuba. Agric Hum Values 26:233–243. doi:10.1007/s10460-008-9156-7

Pinon JR, Muse RL (2010) Coping with the next oil spill: why U.S.-Cuba environmental cooperation is critical. U.S. -Cuba relations at brookings, Issue Brief No 2. http://www.edf.org/documents/11073_brookings-us-cuba.pdf

Pounds JA, Coloma LA (2008) Beware the lone killer. Nat Rep 2:57–59

Rosset PM (1997) Alternative agriculture and crisis in Cuba. IEEE Technol Soc Mag 16(2):19–25

UNEP (2010) Latin America and the Caribbean: environment outlook. Panama City, Panama

UNESCO Science Report (2010) UNESCO Publishing accessible here: http://unesdoc.unesco.org/images/0018/001899/189958e.pdf

United States State Department – Bureau of Western Hemisphere Affairs (2010), Background Note: Cuba. http://www.state.gov/r/pa/ei/bgn/2886.htm

VISIT FLORIDA(R) Research (2010) Calendar year visitor numbers for 2010. http://media.visitflorida.org/research.php

Warwick H (2000) Cuba's organic revolution. Third World Resurgence, 118–119

Warwick H (2001) Cuba's organic revolution. Forum Appl Res Public Policy 16:54–58

Whittle D, Lindeman K (2004) Protecting coastal resources in Cuba: a look at current laws and institutions. In: Proceedings of the coastal society 19th international conference proceedings, Rhode Island

Whittle D, Santos R (2006) Protecting Cuba's environment: efforts to design and implement effective environmental laws and policies in Cuba, in Cuban Studies 37: 73-103, University of Pittsburgh Press, DOI: 10.1353/cub.2007.0018

Whittle D, Lindeman K, Tripp J (2003) International tourism and protection of Cuba's coastal and marine environments. Tulane Environ Law J 16:533–590

Index

J. Gebelein, *A Geographic Perspective of Cuban Landscapes*, Landscape Series 15, DOI 10.1007/978-94-007-2406-8,